HUMPHRY DAVY

MISCELLANEOUS PAPERS

AND

RESEARCHES

ESPECIALLY ON

THE SAFETY LAMP, AND FLAME AND ON THE

PROTECTION OF THE COPPER SHEATHING OF SHIPS

FROM 1815 TO 1828

Elibron Classics
www.elibron.com

Elibron Classics series.

ISBN 1-4212-0362-6 (paperback)
ISBN 1-4212-0361-8 (hardcover)

This Elibron Classics Replica Edition is an unabridged facsimile of the edition published in 1840 by Smith, Elder and Co., London.

MISCELLANEOUS PAPERS

AND

RESEARCHES,

ESPECIALLY ON

THE SAFETY LAMP, AND FLAME,

AND ON THE PROTECTION OF THE

COPPER SHEATHING OF SHIPS,

FROM 1815 TO 1828.

LONDON:

SMITH, ELDER AND CO. CORNHILL.

1840.

[In this volume, the author's later researches are collected, comprising a period of about thirteen years. During this time, his life was more varied than it had ever before been; and his labours of research exhibit the same character. The majority of them were associated with some peculiarity of place, or event, which exciting his curiosity, led, as was his habit, to original inquiry; and in this respect they remarkably display the workings and powers of his mind, and the delight he took in the application of science to useful purposes.

Considering the vast importance of the Author's Researches on Fire-damp, in connexion with the discovery of the Safety-lamp, and the very interesting nature of his Researches on Flame, which the former led to, it has appeared advisable to give ample details on both,—with the hope that they will be acceptable to the scientific reader, and that they will prove useful to practical men, and the more effectually serve the cause of humanity.

The course of publication which the author adopted on this occasion has been noticed in the preliminary volume,—how in the first instance he communicated the results of his inquiries when in progress to the Royal Society,—and how, when completed, he gave a connected account of them, which he published apart. These two plans will here be combined, with this difference only in the execution, that instead of extracts from the original papers which he inserted in the connected account, the papers themselves will be given entire.]

CONTENTS.

I.

ON THE SAFETY LAMP FOR PREVENTING EXPLOSIONS IN MINES, HOUSES LIGHTED BY GAS, SPIRIT WAREHOUSES, OR MAGAZINES IN SHIPS, ETC., WITH SOME RESEARCHES ON FLAME.

DIRECTIONS FOR THE PLATES.

ON THE SAFETY LAMP

FOR

PREVENTING EXPLOSIONS IN MINES,

HOUSES LIGHTED BY GAS, SPIRIT WAREHOUSES,

OR

MAGAZINES IN SHIPS, ETC.

WITH SOME

RESEARCHES ON FLAME.

ADVERTISEMENT.

This work was published in 1818; but a part of the edition having remained unsold, and the recent occurrence of some severe accidents from explosions in coal-mines, and houses lighted by gas, and the loss of ships by fire, having shown that the precautions which it was intended to describe, either are not known, or are not attended to, I have thought it might assist the cause of humanity to advertise the book a second time. I have added to it a few additional paragraphs, which contain some new facts, and some practical results, connected with the use of the Safety Lamp; most of the last occurred to me during journies that I have made, for the purpose of introducing this invention into the principal mines of Europe, in which inflammable air is found.

Park Street,
March 20, 1825.

I.

ON THE SAFETY LAMP; WITH SOME RESEARCHES ON FLAME.

PREFACE.

I HAVE thought it right to collect and to publish in a connected form an account of all the researches that I have made on the subject of explosions from inflammable air, and the modes in which they may be prevented, as well as the collateral investigations to which they have given rise, with the hope of presenting a permanent record on this important subject to the practical miner, and of enabling the friends of humanity to estimate and apply those resources of science, by which a great and permanently existing evil may be subdued.

In connecting general views on the subject with extracts from papers published in the Transactions of the Royal Society, I fear I shall sometimes be accused of repetitions; but in a case where human life is concerned, and by which human happiness may be deeply affected, I shall not dread the accusation of dwelling too long upon, or treating too often of, precautionary measures.*

* [These measures, it is to be regretted, have been too little studied by many of those persons most deeply concerned in them. In proof, it may be mentioned that the Author's work on the Safety Lamp, expressly designed for practical purposes,—though published now twenty-two years and in a cheap form,—was in so little request, that a second edition of it was never required; and that in the parliamentary inquiry instituted in 1835, relative to explosions in mines, some of the witnesses who gave evidence before the Select Committee of the House of

The names of three persons will be found mentioned in these pages as having assisted in the investigations. The public owe much to the Rev. John Hodgson and to Mr. Buddle for having been the first persons to make experiments upon the Safety Lamp in explosive atmospheres in the mine, and for elucidating its practical application and rendering it familiar to the miner; and I am myself indebted to Mr. Michael Faraday for much able assistance in the prosecution of my experiments.

I have given the extracts from my papers nearly in the order in which they were published; which will, I hope, both render the facts more intelligible, and show the gradual progress of the inquiry: in which every step was furnished by experiments or induction, in which nothing can be said to be owing to accident, and in which the most simple and useful combination arose out of the most complicated circumstances.

The result of these labours will, I trust, be useful to the cause of science, by proving that even the most apparently abstract philosophical truths may be connected with applications to the common wants and purposes of life.

The gratification of the love of knowledge is delightful to every refined mind; but a much higher motive is offered for indulging in it, when that knowledge is felt to be practical power, and when that power may be applied to lessen the miseries or increase the comforts of our fellow-creatures.

London, May 14, 1818.

Commons, men having highly responsible situations in collieries, even in fiery ones, where the safety lamp was in use, did not appear to be acquainted with this publication, and expressed surprise at hearing statements from Mr. Buddle relative to precautions in using the lamp in situations of danger, which are distinctly pointed out in it.]

I.—GENERAL VIEWS OF THE PROGRESS OF THE RESEARCHES ON THE SAFETY LAMP, AND OF THE PRINCIPLES ON WHICH ITS SECURITY DEPENDS.

The use of pit-coal in Britain is connected, not only with the necessaries, comforts, and enjoyments of life, but also with the extension of our most important arts, our manufactures, commerce, and national riches.

Essential in affording warmth and preparing food, it yields a sort of artificial sunshine, and in some measure compensates for the disadvantages of our climate. By means of it, metallurgical processes are carried on, and the most important materials of civilized life furnished, the agriculturist is supplied with an useful manure, and the architect with a necessary cement. Not only manufactories and private houses, but even whole streets are lighted by its application; and, in furnishing the element of activity in the steam-engine, it has given a wonderful impulse to mechanical and chemical ingenuity, diminished to a great extent human labour, and increased, in a high degree, the strength and wealth of the country.

Every thing connected with the permanent supply of such a material, is worthy of scientific consideration; and to remove obstacles, difficulties, or dangers connected with its production, is not unimportant to the State.

Since the earliest period of the application of mineral coal* to the purposes of fuel, the explosions in coal-

* Coal was certainly worked in the neighbourhood of Newcastle in 1245. See Brande's History of Newcastle, vol. ii. p. 253.

mines from inflammable air,* or fire-damp, have been regarded as the greatest evil occurring in the working of the mines. The strata of coal lie usually parallel, or nearly parallel to the surface, at certain depths beneath it; and this coal, when the pressure of the superincumbent material is removed, affords inflammable air; which is disengaged, not only in the common operations of mining, when the coal is broken and removed, but is likewise permanently evolved, often in enormous quantities,† from fissures in the strata.‡

When it has accumulated in any part of the gallery or chamber of a mine, so as to be mixed in certain proportions with common air, the presence of a lighted candle or lamp causes it to explode, and to destroy, injure, or burn whatever is exposed to its violence.

To give detailed accounts of the tremendous accidents, owing to this cause, would be merely to multiply pictures of death, and of human misery. The phenomena are always of the same kind. The miners are either immediately destroyed by the explosion, and

* Called Grisoux in Flanders, and Feu Brisson in the southern departments of France.

† This phenomenon is called, in the language of the North country miners, a *blower*.

‡ [In bituminous coal, it is most abundant. In the coal, previous to extrication, there is reason to believe that it is always condensed; and occasionally probably to such a degree, as to be liquid; Mr. Smith's account of the phenomena he witnessed in a coal-mine in Novia Scotia, as given by him before the Select Committee on Accidents in Mines, is very accordant with such a supposition. "As soon as the coal was struck at the depth of 180 feet, it appeared to throw the whole coal-mine into a regular state of mineral fermentation. The gas roared as the miner struck the coal with his pick; it would often go off like the report of a pistol; and at times I have seen it burst pieces of coal off the solid wall. The noise which the gas and water made in issuing from the coal, was like a hundred thousand snakes hissing at each other."—Minutes of Evidence, p. 276.]

thrown with the horses and machinery through the shaft into the air, the mine becoming as it were an enormous piece of artillery, from which they are projected; or they are gradually suffocated, and undergo a more painful death, from the carbonic acid and azote remaining in the mine, after the inflammation of the fire-damp; or what, though it appears the mildest, is perhaps the most severe fate, they are burnt or maimed, and often rendered incapable of labour, and of healthy enjoyment for life.

The fire-damp is found in the greatest quantity, and is most dangerous in the deepest mines; but it likewise often occurs in superficial excavations; and I have now a letter, of the date of June 8, 1816, in my possession, in which it is stated that in the very commencement of working a coal-mine in Shropshire, several miners were killed, and others severely burnt.

Modes of preventing accidents from fire-damp, have been ardently sought for by all persons connected with coal mines, and it has even occupied the attention of an enlightened government. In consequence of some explosions which prevented the miners from working the coal mines at Briançon, in Dauphiny,* the Duke de Choiseul, at that time Prime Minister of France, recommended the subject to the consideration of the Academy of Sciences, and a Committee was appointed, who made it for some time the object of their attention; but the plan that they proposed for avoiding the danger, was a common mode of ventilation.

This evil of the fire-damp, though belonging to all coal mines, had been most severely experienced in those of Hainault, in Flanders, and the infinitely more

* Histoire de l'Academie Royale, 1763, p. 1.

important mines in the neighbourhood of Whitehaven and Newcastle, in this country.

The number of dreadful accidents, indeed, which had happened within the last three or four years in the last mentioned districts, particularly that by which ninety-six persons were destroyed in the Felling colliery, had so strongly impressed the minds of a number of benevolent persons belonging to or connected with the coal districts, that it was said to be in their contemplation to bring the subject before Parliament, that by making it a national question, it might obtain that consideration which its importance demanded.

When I first turned my attention particularly to the subject, which was in August, 1815, in consequence of a letter from the Rev. Dr. Gray, there appeared very little hope of finding an efficacious remedy. The resources of modern chemical science had been fully applied in ventilation, in the improved plans of Mr. Buddle; the comparative lightness of the fire-damp was well understood, every precaution was taken to preserve the communications open; and the currents of air were promoted or occasioned, not only by furnaces, but likewise by air-pumps and steam apparatus.

Sir James Lowther had observed early in the last century that the fire-damp in its usual form was not inflammable by sparks from flint and steel; and a person in his employment had invented a mill for giving light by the collision of flint and steel,* and this was the only instrument except common candles employed in the dangerous parts of the British collieries. Yet instances of explosion have been known

* Said to be Mr. Spedding, in Hutchinson's History of Cumberland, Article, Whitehaven.

from the steel-mill, and it required manual labour for its use. In Flanders, amadou or fungus tinder had been occasionally employed in dangerous parts of the mine, but the light yielded by this substance was much too feeble to be used for working the mines, and only enabled the miners to find their way for particular occasions.

M. de Humboldt, the justly celebrated philosophical traveller, in 1796 conceived and executed the plan of a lamp* for giving light in mines where a common candle would not burn or produce explosion; but it was founded on the principle of entire insulation from the air, and could burn only for a short time till the air contained within it was exhausted. A lamp upon a plan similar as to insulation, was contrived by Dr. Clanny, in 1813, but he supplied his light with air from the mine through water by bellows, and it went out in explosive atmospheres, and to be employed required to be worked by hand, or by machinery; and neither M. de Humboldt's lamps, or Dr. Clanny's, had, for obvious reasons I believe, ever been used in coal-mining.

The great object, one rather to be ardently desired than confidently expected, was to find a light, which, at the same time that it enabled the miner to work with security in explosive atmospheres, should likewise consume the fire-damp. Having learnt from Mr. Buddle the degree of light required for the common operations of the workmen, I made several experiments with the hope of obtaining such a light without active inflammation. I tried Kunckel's, Canton's, and Baldwin's phosphorus, and likewise the electrical light in close vessels, but without success. I had a lamp made

* Journal des Mines, tom. viii. p. 829.

with two valves, which closed in atmospheres contaminated with fire-damp, by the increased heat of the flame produced by the combustion of the gas, but this lamp could not be used in an explosive atmosphere.

It will be unnecessary to dwell upon preliminary and unsuccessful attempts, and I shall proceed to describe the origin and progress of those investigations which led me to the discovery of the principles by which explosion and flame may be arrested and regulated; and by means of which, the miner is not only able to subdue and control, but likewise to render useful his most dangerous enemy.

I first began with a minute chemical examination of the substance with which I had to contend. The analysis of various specimens of fire-damp shewed me that the pure inflammable part of it was light carburetted hydrogen, as Dr. Henry had before stated, hydrogen or pure inflammable air combined with charcoal or carbon.

I made numerous experiments on the circumstances under which it explodes, and the degree of its inflammability. I found that it required to be mixed with very large quantities of atmospheric air to produce explosion; even when mixed with three or nearly four times its bulk of air, it burnt quietly in the atmosphere, and extinguished a taper. When mixed with between five and six times its volume of air it exploded feebly: it exploded with most energy when mixed with seven or eight times its volume of air, and mixtures of fire-damp and air retained their explosive power when the proportions were one of gas to fourteen of air. When the air was in larger quantity the flame of a taper was merely enlarged in the mixture, an effect which was still perceived in thirty parts of air to one of gas.

I found the fire-damp much less combustible than other inflammable gases. It was not exploded or fired by red-hot charcoal or red-hot iron; it required iron to be white-hot, and itself in brilliant combustion, for its inflammation. The heat produced by it in combustion was likewise much less than that of most other inflammable gases, and hence, in its explosion, there was much less comparative expansion.

On mixing 1 part of carbonic acid or fixed air with 7 parts of an explosive mixture of fire-damp, or 1 part of azote with 6 parts, their powers of exploding were destroyed.

In exploding a mixture in a glass tube of one-fourth of an inch in diameter and a foot long, more than a second was required before the flame reached from one end to the other: and I found that in tubes of one-seventh of an inch in diameter, explosive mixtures could not be fired when they were opened in the atmosphere; and that metallic tubes prevented explosion better than glass tubes.

In reasoning upon these various phenomena it occurred to me, as a *considerable* heat was required for the inflammation of the fire-damp, and as it produced in burning comparatively a *small degree* of heat, that the effect of carbonic acid and azote, and of the surfaces of small tubes in preventing its explosion, depended upon their cooling powers; upon their lowering the temperature of the exploding mixture so much that it was no longer sufficient for its continuous inflammation.

This idea, which was confirmed by various obvious considerations, led to an immediate result—the possibility of constructing a lamp, in which the cooling powers of the azote or carbonic acid, formed by combustion or the cooling powers of the apertures, through which the air entered or made its exit, should prevent the communication of explosion.

I first tried the effects of lamps in which there was a very limited circulation of air; and I found that when a taper in a close lantern was supplied with air so as to burn feebly from very small apertures below the flame, and at a considerable distance from it, it became extinguished in explosive mixtures; but I ascertained that precautions which it would be dangerous to trust to workmen were required to make this form of a lamp safe, and that at best it could give only a feeble light; and I immediately adopted systems of tubes above and below, of that diameter in which I had ascertained that explosions would not take place.

In trying my first tube lamp in an explosive mixture I found that it was safe; but unless the tubes were very short and numerous, the flame could not be well supported; and in trying tubes of the diameter of one-seventh or one-eighth of an inch I determined that they were safe only to small quantities of explosive mixture, and when of a given length; and that tubes even of a much smaller diameter communicated explosion from a close vessel. Hence I took a new method of ascertaining the safety of my apertures, and of trying different forms of apertures.

I had a vessel furnished with wires by which the electrical spark could be taken in an explosive mixture, and which was larger in capacity than a safe lamp or lantern was required to be. I placed my flame sieves, i. e. my systems of apertures, between this jar and a bladder containing likewise an explosive mixture, and I judged the aperture to be safe only when they stopped explosion acting upon them in this concentrated way.

In this mode of experimenting I soon discovered that

a *few apertures* even of very small diameter were not safe unless their sides were very deep; that a single tube of one-twenty-eighth of an inch in diameter and two inches long suffered the explosion to pass through it; and that a *great number* of small tubes or of apertures, stopped explosion even when the depths of their sides was only equal to their diameters; and at last I arrived at the conclusion that a *metallic tissue,* however thin and fine, of which the apertures filled more space than the cooling surface, so as to be permeable to air and light, offered a perfect barrier to explosion, from the force being divided between, and the heat communicated to an immense number of surfaces.

My first safety lamps, constructed on these principles, gave light in explosive mixtures containing a great excess of air, but became extinguished in explosive mixtures in which the fire damp was in sufficient quantity to absorb the whole of the oxygen of the air, so that such mixtures never burnt continuously at the air feeders, which in lamps of this construction was important, as the increase of heat, where there was only a small cooling surface, would have altered the conditions of security.

I made several attempts to construct safety lamps which should give light in all explosive mixtures of fire-damp, and after complicated combinations I at length arrived at one evidently the most simple, that of surrounding the light entirely by wire gauze, and making the same tissue feed the flame with air and emit light.

In plunging a light surrounded by a cylinder of fine wire-gauze into an explosive mixture I saw the whole cylinder become quietly and gradually filled with flame; the upper part of it soon appeared red hot; yet *no* explosion was produced.

It was easy at once to see that by increasing the cooling surface in the top, or in any other part of the lamp, the heat acquired by it might be diminished to any extent; and I immediately made a number of experiments to perfect this *invention,* which was evidently the one to be adopted, as it excluded the necessity of using glass or any fusible or brittle substance in the lamp, and not only deprived the fire damp of its explosive powers, but rendered it an useful light.

Though all the specimens of fire damp which I had examined consisted of carburetted hydrogen mixed with different small proportions of carbonic acid and common air, yet some phenomena that I observed in the combustion of a blower, induced me to believe that small quantities of olefiant gas may be sometimes evolved in coal mines with the carburetted hydrogen. I therefore resolved to make all lamps safe to the test of the gas produced by the distillation of coal, which when it has not been exposed to water always contains olefiant gas. I placed my lighted lamps in a large glass receiver, through which there was a current of atmospherical air, and by means of a gasometer filled with coal gas, I made the current of air which passed into the lamp more or less explosive, and caused it to change rapidly or slowly at pleasure, so as to produce all possible varieties of inflammable and explosive mixtures; and I found that iron wire-gauze, composed of wires from one-fortieth to one-sixtieth of an inch in diameter, and containing twenty-eight wires or 784 apertures to the inch, was safe under all circumstances in atmospheres of this kind: and I consequently adopted this material in guarding lamps for the coal mines, where in January, 1816, they were immediately adopted, and have long been in general use.

Observations upon them in their working state, and upon the circumstances to which they are exposed, have led to a few improvements or alterations, merely connected with the modes of increasing light or diminishing heat, which were very obvious from the original construction; and experiments on the nature of flame, and the modifications of combustion led me, in January, 1817, to an important practical addition, founded entirely upon a new principle.

The wire-gauze lamp, in its common form, burns in all atmospheres that are explosive; and, by suspending or placing in it a little cage of wire of platinum or palladium from one-sixtieth to one-seventieth of an inch in thickness, it yields a light in atmospheres too much contaminated with fire-damp to be explosive, a slow combination being occasioned by the heated platinum between the elements of the gas and oxygen, which produces sufficient heat to keep the metals of low conducting power and low capacity for heat permanently ignited, wherever there is air enough to support life without suffering.

I shall conclude this view, by some general observations on flame and combustion, which will show more distinctly the causes and the limits of safety in lamps, and which will demonstrate the danger of combinations made with an imperfect knowledge of the principles of security.

Flame may be defined to be aëriform, or gaseous matter, heated to such a degree as to be luminous, and it may be produced independent of any chemical changes, as is shown in the discharge of voltaic electricity, through an undecomposable gas. Very concentrated electricity, in passing through bodies, constantly heats them, whether they be solid, fluid, or gaseous:

and by the voltaic apparatus, the nature of flame is distinctly shown. Flames are conical, because the greatest heat is in the centre of the mass; and because heated air rapidly ascends through cooler air. When *flame* is produced in chemical combination, gaseous matter is the cause of it; and the heat of flames seems always proportional, other circumstances remaining the same, to the rapidity of combination, and to the density of the gases combining. Thus, the heat of flames diminishes by rarefaction, and increases by condensation. Whenever combustible gaseous matter burns in the atmosphere, it must first mix with a certain quantity of air: if it require a high temperature for its combustion, it will be easily extinguished by rarefaction, or by cooling agencies, whether of solid surfaces, or mixtures of incombustible gases: if it require a very low temperature for its combustion, it will burn in highly rarefied air, or under considerable cooling agencies.

By heating strongly gases that burn with difficulty, their continued inflammation becomes easy, in consequence of increments of heat occasioned by combustion of small quantities, which, under any other circumstances, would not produce continued combustion. Hence, if mixtures of fire-damp are burnt from systems of tubes or canals, or metallic plates, which have small, radiating, and cooling surfaces: though these systems are safe at first, they become dangerous, as they are heated.* Where currents are occasioned which concentrate explosive mixtures by the air feeders in lamps being below,* and made in thick metallic plates or

* I warn the coal miner against any pretended safety lamps made in this manner, and which, to superficial observers, may appear to be constructed upon principles of security, but in which these principles cannot really exist.

canals, there being an increment of heat within, and a very small radiating surface without, as the heat increases, the combustion of the explosive mixture will gradually extend further, and at last communicate with the external air; for explosion will be communicated by any aperture, however small, provided it be sufficiently heated. This circumstance is shown in a very elegant manner, in burning concentrated mixtures of oxygen and hydrogen at the end of a long tube of one-sixtieth or one-seventieth of an inch in diameter, when the experiment begins (the tube being cool) there is no danger; gradually, however, as it becomes heated, the combustion steals as it were down the tube, and at last reaches the reservoir of the gases.

Where one set of air-feeders only are attempted in a lamp, they should present an uniform surface, so that the radiating powers of the metal, and the cooling powers of the external gases, may immediately balance the heating powers of the internal gases; and where the radiating tissue is connected with other parts of the lamp, these parts should be so massy as to be slightly heated only, and present no means for a gradual accumulation of heat. Wire gauze, as it offers a greater extent of radiating surface than perforated metallic plates, is the best material for the guard of lamps; and by being made of the proper degree of fineness, it will form a barrier for every species of explosion requiring temperatures higher than those of our atmosphere: but the apertures must be smaller, and the radiating surfaces greater, in proportion to the inflammability of the gas: and currents of concentrated explosive mixtures, acting even for any length of time, may be stopped by reduplications of wire gauze. Wire gauze for lamps must not be made of, or covered with any easily-combustible metal: fine brass wire is improper, on account of the zinc it contains; and the iron wire should not be tinned. The body of the lamp should be of copper, riveted together, or of massy cast-brass or cast-iron; the screws should fit tight; no aperture, *however small*, should be suffered to exist in the body of the lamp; and the trimming wire should move through a long tight tube.

Flame, whether produced by the combustion of large or small quantities of explosive mixture, may be always extinguished or destroyed by certain cooling agencies; and, in proportion to the heat required to carry on the combustion, so is it more easily destroyed. The temperature of metal, even when white-hot, is far below that of flame; and hence red-hot gauze, in sufficient quantity, and of the proper degree of fineness, will abstract sufficient heat from the flame of carburetted hydrogen or fire-damp, to extinguish it.

Combinations of gases may be occasioned by a heat not sufficient to raise their temperature into flame, but they still produce heat during their combination, as is evident from what has been stated, page 15: and when in a mixture containing air and combustible gases, the cooling agencies are too great to permit the appearance of flame, or their continued combination; still this combination may be kept up by the ignition of platinum,—so that with a certain quantity of platinum in a cage of wire gauze, the fire-damp may be entirely consumed without flame, yielding only a beautiful light by the ignition of solid matter.

II.—PAPERS PUBLISHED IN THE PHILOSOPHICAL TRANSACTIONS, AND IN THE JOURNAL OF SCIENCE AND THE ARTS, ON THE FIRE-DAMP, THE SAFETY LAMP, AND ON FLAME.

1.—*On the Fire-damp of Coal Mines, and on the Methods of Lighting the Mines, so as to prevent its Explosion.**

THE accidents arising from the explosion of the fire-damp, or inflammable gas of coal-mines, mixed with atmospherical air, are annually becoming more frequent and more destructive in the collieries in the North of England.

A Committee has been, for some time, formed at Sunderland, for the benevolent purpose of investigating the causes of these accidents, and of searching for means of preventing them. In consequence of an invitation from the Rev. Dr. Gray, [late Bishop of Bristol], one of the most active members of this Committee, I was induced to turn my attention to the subject. I went to the North of England, and visited some of the principal collieries in the neighbourhood of Newcastle, for the purpose of ascertaining the conditions of the workings, and the state of their ventilation. I found the greatest desire to assist my inquiries with the gentlemen acquainted with the northern collieries, as well as with inspectors or viewers of the mines; and I have particular obligations on this point to the Rev. Dr. Gray, Cuthbert Ellison, Esq. M.P., the Rev. John Hodgson, Mr. Buddle, and Mr. Dunn. Dr. Fenwick and Dr. Clanny, and Mr. Fenwick, likewise kindly offered me their assistance.

* [From the Philosophical Transactions for 1816, read before the Royal Society, Nov. 9, 1815.]

From the information which I collected on the spot, increased by the perusal of a report of Mr. Buddle on the state of the mines, I was convinced that, as far as ventilation was concerned, the resources of modern science had been fully employed; and that a mode of preventing accidents was only to be sought for, in a method of lighting the mines free from danger, and which, by indicating the state of the air in the part of the mine where inflammable air was disengaged, so as to render the atmosphere explosive, should oblige the miners to retire till the workings were properly cleared.

An account of an ingenious apparatus for burning a candle, supplied with atmospherical air, by a bellows through water, has been published in the Philosophical Transactions by Dr. Clanny; but I believe this apparatus has not yet been used in any other collieries.

The common means employed for lighting those parts of the mine where danger is apprehended from the fire-damp, is by a steel wheel, which, being made to revolve in contact with flint, affords a succession of sparks; but this apparatus always requires a person to work it; and though much less liable to explode the fire-damp than a common candle, yet it is said to be not entirely free from danger.

Mr. Buddle having obligingly shown to me the degree of light required for working the collieries, I made several experiments, with the hope of producing such a degree of light, without active inflammation; I tried Kunckel's, Canton's, and Baldwin's phosphorus, and likewise the electrical light in close vessels, but without success; and even had these degrees of light been sufficient, the processes for obtaining them, I found, would be too complicated and difficult for the miners.

The fire-damp has been shewn by Dr. Henry, in a

very ingenious paper,* published in the nineteenth volume of Nicholson's Journal, to be light carburetted hydrogen gas, and Dr. Thomson has made some experiments upon it; but the degree of its combustibility, as compared with that of other inflammable gases, has not, I believe, been examined, nor have many different specimens of it been analyzed; and it appears to me, that some minute chemical experiments on its properties ought to be preliminary steps to inquiries respecting methods of preventing its explosion. I therefore procured various specimens of the fire-damp in its purest state, and made a number of experiments upon it. And in examining its relations to combustion, I was so fortunate as to discover some properties belonging to it, which appear to lead to very simple methods of lighting the mines, without danger to the owners, and which, I hope, will supply the desideratum so long anxiously required by humanity. I shall, in the following pages, have the honour of describing these properties, and the methods founded upon them, to the Royal Society, and I shall conclude with some general observations.

The fire damp is produced in small quantities in coal mines, during the common process of working.

The Rev. Mr. Hodgson informed me, that on pounding some common Newcastle coal fresh from the mine in a cask furnished with a small aperture, the gas from the aperture was inflammable. And on breaking some large lumps of coal under water, I ascertained that they

* [This paper was published in 1808. The conclusion Dr. Henry arrived at was the following: "It is to be regretted, that the analysis of the fire-damp affords no encouragement to expect that it can ever be destroyed in coal-mines by any chemical process, as has lately been proposed. The only feasible method of preventing the dreadful consequences of its combustion is, to enforce the steady execution of a well planned system of ventilation."]

gave off inflammable gas.* Gas is likewise disengaged from bituminous schist, when it is worked.

The great sources of the fire-damp in mines, are, however, what are called blowers, or fissures in the broken strata, near dykes, from which currents of fire-damp issue in considerable quantity, and sometimes for a long course of years.† When old workings are broken into, likewise, they are often found filled with fire-damp; and the deeper the mine the more common in general is this substance.

I have analysed several specimens of the fire-damp in the laboratory of the Royal Institution; the pure inflammable part was the same in all of them, but it was sometimes mixed with small quantities of atmospherical air, and in some instances with azote and carbonic acid.

* This is probably owing to the coal strata having been formed under a pressure greater than that of the atmosphere, so that they give off elastic fluid when they are exposed to the free atmosphere: and probably coals containing animal remains, evolve not only the fire-damp, but likewise azote and carbonic acid.

In the Apennines, near Pietra Mala, I examined a fire produced by gaseous matter, constantly disengaged from a schist stratum: and from the results of combustion, I have no doubt but that it was pure fire-damp. Mr. M. Faraday, who accompanied me, and assisted me in my chemical experiments, in my journey, collected some gas from a cavity in the earth about a mile from Pietra Mala, then filled with water, and which, from the quantity of gas disengaged, is called Aqua Buja. I analysed it in the Grand Duke's laboratory at Florence, and found that it was pure light carburetted hydrogen, requiring two volumes of oxygen for its combustion, and producing a volume of carbonic acid gas.

It is very probable, that these gases are disengaged from coal strata beneath the surface, or from bituminous schist above coal; and at some future period new sources of riches may be opened to Tuscany from this invaluable mineral treasure, the use of which in this country has supplied such extraordinary resources to industry.

† Sir James Lowther found a uniform current produced in one of his mines for two years and nine months. Phil. Trans. vol. xxxviii. p. 112.

Of six specimens collected by Mr. Dunn from a blower in the Hepburn Colliery, by emptying bottles of water close to it, the purest contained one-fifteenth only of atmospherical air, with no other contamination; and the most impure contained five-twelfths of atmospherical air; so that this air was probably derived from the circumambient air of the mine. The weight of the purest specimen was for 100 cubical inches 19·5 grains.

One measure of it required for its complete combustion by the electric spark nearly two measures of oxygen, and they formed nearly one measure of carbonic acid.

Sulphur heated strongly and repeatedly sublimed in a portion of it freed from oxygen by phosphorus, produced a considerable enlargement of its volume, sulphuretted hydrogen was formed, and charcoal precipitated; and it was found that the volume of the sulphuretted hydrogen produced, when it was absorbed by solution of potassa, was exactly double that of the fire-damp decomposed.

It did not act upon chlorine in the cold; but, when an electric spark was passed through a mixture of 1 part of it with 2 of chlorine, there was an explosion, with a diminution to less than one-fourth, and much charcoal was deposited.

The analysis of specimens of gas sent to my friend John George Children, Esq. by Dr. Clanny, afforded me similar results; but they contained variable quantities of carbonic acid gas and azote.

Different specimens of these gases were tried by the test of exposure to chlorine both in darkness and light: they exhibited no marks of the presence of olefiant gas or hydrogen; and the residuum produced by detonation with chlorine showed them to be free from carbonic oxide.

It is evident, then, that the opinion formed by other chemists respecting the fire-damp is perfectly correct; and that it is the same substance as the inflammable gas of marshes, the exact chemical nature of which was first demonstrated by Mr. Dalton; and that it consists, according to my view of definite proportions, of 4 proportions of hydrogen in weight 4, and 1 proportion of charcoal in weight 11·5.

I made several experiments on the combustibility and explosive nature of the fire-damp. When 1 part of fire-damp was mixed with 1 of air, they burnt by the approach of a lighted taper, but did not explode; 2 of air and 3 of air to 1 of gas produced similar results. When 4 of air and 1 of gas were exposed to a lighted candle, the mixture being in the quantity of 6 or 7 cubical inches in a narrow-necked bottle, a flame descended through the mixture, but there was no noise: 1 part of gas inflamed with 6 parts of air in a similar bottle, produced a slight whistling sound: 1 part of gas with 8 parts of air, rather a louder sound: 1 part with 10, 11, 12, 13, and 14 parts, still inflamed, but the violence of combustion diminished. In 1 part of gas and 15 parts of air, the candle burnt without explosion with a greatly enlarged flame; and the effect of enlarging the flame, but in a gradually diminished ratio, was produced as far as 30 parts of air to 1 of gas.

The mixture which seemed to possess the greatest explosive power, was that of 7 or 8 parts of air to 1 of gas; but the report produced by 50 cubical inches of this mixture was less than that produced by one-tenth of the quantity of a mixture of 2 parts of atmospherical air and 1 of hydrogen.

It was very important to ascertain the degree of heat required to explode the fire-damp mixed with its proper proportion of air.

I found that a common electrical spark would not explode 5 parts of air and 1 of fire-damp, though it exploded 6 parts of air and 1 of damp: but very strong sparks from the discharge of a Leyden jar, seemed to have the same power of exploding different mixtures of the gas as the flame of the taper. Well burned charcoal, ignited to the strongest red heat, did not explode any mixture of air and of the fire-damp; and a fire made of well burned charcoal, i. e. charcoal that burned without flame, was blown up to whiteness by an explosive mixture containing the fire-damp, without producing its inflammation. An iron rod at the highest degree of red heat, and at the common degree of white heat, did not inflame explosive mixtures of the fire-damp; but, when in brilliant combustion, it produced the effect.

The flame of gaseous oxide of carbon as well as of olefiant gas exploded the mixtures of the fire-damp.

In respect of combustibility, then, the fire-damp differs most materially from the other common inflammable gases. Olefiant gas, which I have found explodes mixed in the same proportion with air, is fired by both charcoal and iron heated to redness. Gaseous oxide of carbon, which explodes when mixed with two parts of air, is likewise inflammable by red-hot iron and charcoal. And hydrogen, which explodes when mixed with three-sevenths of its volume of air, takes fire at the lowest visible heat of iron and charcoal; and the case is the same with sulphuretted hydrogen.

I endeavoured to ascertain the degree of expansion of mixtures of fire-damp and air during their explosion, and likewise their power of communicating flame through apertures to other explosive mixtures.

I found that when 6 of air and 1 of fire-damp were exploded over water by a strong electrical spark, the explosion was not very strong, and, at the moment of the greatest expansion, the volume of the gas did not appear to be increased more than one-half.*

In exploding a mixture of 1 part of gas from the distillation of coal, and 8 parts of air in a tube of a quarter of an inch in diameter, and a foot long, more than a second was required before the flame reached from one end of the tube to the other; and I could not make any mixture explode in a glass tube one-seventh of an inch in diameter: and this gas was more inflammable than the fire-damp, as it consisted of carburetted hydrogen gas mixed with some olefiant gas.

In exploding mixtures of fire-damp and air in a jar connected with the atmosphere by an aperture of half an inch, and connected with a bladder by a stop-cock, having an aperture of about one-sixth of an inch,† I found that the flame passed into the atmosphere, but did not communicate through the stop-cock, so as to inflame the mixture in the bladder: and in comparing the power of tubes of metal and those of glass, it appeared that the flame passed more readily through glass tubes of the same diameter; and that explosions were stopped by metallic tubes of one-fifth of an inch,‡ when

* This appears the expansion when the tube is very small; in larger tubes, it is considerably more. The volume of the gas appears at least tripled during the explosion.

† Since these experiments were made, Dr. Wollaston has informed me, that he and Mr. Tennant had observed some time ago, that mixtures of the gas from the distillation of coal and air, would not explode in very small tubes.

‡ I do not give this result as perfectly exact, as the base of the metallic tube had not the same polish as that of the tube of glass.

they were an inch and a half long; and this phenomenon probably depends upon the heat lost during the explosion in contact with so great a cooling surface, which brings the temperature of the first portions exploded below that required for the firing of the other portions. Metal is a better conductor of heat than glass: and it has been already shown that the fire-damp requires a very strong heat for its inflammation.

Mixture of the gas with air I found, likewise, would not explode in metallic canals or troughs, when their diameter was less than the one-seventh of an inch, and their depth considerable in proportion to their diameter; nor could explosions be made to pass through such canals.

Explosions likewise I found would not pass through very fine wire sieves or wire gauze.

I mixed azote and carbonic acid in different quantities with explosive mixtures of fire-damp, and I found that even in very small proportions they diminished the velocity of the inflammation. Azote, when mixed in the proportion of 1 to 6 of an explosive mixture, containing 12 of air and 1 of fire-damp, deprived it of its power of explosion; when 1 part of azote was mixed with 7 of an explosive mixture, only a feeble blue flame slowly passed through the mixture.

1 part of carbonic acid to 7 of an explosive mixture deprived it of the power of exploding; so that its effects are more remarkable than those of azote; probably, in consequence of its greater capacity for heat, and probably, likewise, of a higher conducting power connected with its greater density.

The consideration of these various facts, led me to adopt a form of a lamp, in which the flame, by being supplied with only a limited quantity of air, should produce such a quantity of azote and carbonic acid, as to prevent the explosion of fire-damp, and which, by the nature of its apertures for giving admittance and exit to the air, should be rendered incapable of communicating any explosion to the external air.

If in a close lantern, supplied with a small aperture below and another above, a lighted lamp having a very small wick, be placed, the natural flame gradually diminishes, till it arrives at a point at which the supply of air is sufficient for the combustion of a certain small quantity of oil; if a lighted taper be introduced into the lantern through a small door in the side, which is instantly closed, both lights will burn for a few seconds, and be extinguished together.

A similar phenomenon occurs, if, in a close lantern, supplied with a quantity of air merely sufficient to support a certain flame, a mixture of fire-damp and air is gradually admitted, the first effect of the fire-damp is to produce a larger flame round that of the lamp, and this flame, consuming the oxygen which ought to be supplied to the flame of the lamp, and the standard of the power of the air to support flame being lowered by the admixture of the fire-damp and by its rarefaction, both the flame of the fire-damp, and that of the taper are extinguished together; and as the air contained a certain quantity of azote and carbonic acid before the admission of the fire-damp, their effect, by mixing with it, is such as to prevent an explosion in any part of the lantern.

I tried several experiments on the burning of a flame in atmospheres containing fire-damp. I enclosed a taper in a little close lantern, having a small aperture below and a larger one above, of such size that the taper burnt with a flame a little below its natural size. I placed this lantern, the taper being lighted, on a stand under a large glass receiver standing in water, having a curved tube containing a little water, adapted to its top to confine the air, and which was of such a capacity as to enable the candle to burn for some minutes; I then rapidly threw a quantity of fire-damp into the receiver from a bladder, so as to make the atmosphere in it explosive. As the fire-damp mixed with the air, the flame of the taper gradually enlarged, till it half filled the lantern; it then rapidly diminished, and was suddenly extinguished without the slightest explosion. I examined the air of the receiver after the experiment, and found it highly explosive.

I tried similar experiments, throwing in mixtures of air and fire-damp, some containing the maximum, and others the minimum of fire-damp necessary for explosion, and always with the same satisfactory results. The flame considerably increased, and was soon extinguished.

I introduced a lighted lantern, to which air was supplied by two glass tubes of one-tenth of an inch in diameter, and half an inch long, into a large jar containing an explosive mixture of one part of fire-damp and ten parts of air; the taper burnt at first with a feeble light, the flame soon became enlarged, and was then extinguished. I repeated these experiments several times, and with a perfect constancy of result.

It is evident, then, that to prevent explosions in coal mines, it is only necessary to use air-tight lanterns, supplied with air from tubes or canals of small diameter, or from apertures covered with wire-gauze placed below the flame, through which explosions cannot be communicated, and having a chimney at the upper part, as a similar system for carrying off the foul air; and common lanterns may be easily adapted to the purpose by being made air-tight in the door and sides, by being furnished with the chimney, and the system of safety apertures below and above.

The principle being known, it is easy to adapt, and multiply practical applications of it.

The first safe lantern that I had constructed, was made of tin plate, and the light emitted through four glass plates in the sides. The air was admitted round the bottom of the flame from a number of metallic tubes of one-third of an inch in diameter, and an inch and a half long. The chimney was composed of two open cones, having a common base perforated with many small apertures, and fastened to the top of the lantern, which was made tight in a pneumatic ring containing a little oil; the upper and lower apertures in the chimney were about one-third of an inch: the lamp, which was fed with oil, gave a steady flame of about an inch high and half an inch in diameter. When the lantern was slowly moved, the lamp continued to burn, but more feebly, and when it was rapidly moved it went out. To obviate this circumstance, I surrounded the bottom of the lantern with a perforated rim; and this arrangement perfectly answered the end proposed. I had another chimney fitted to this lantern furnished with a number of safety tin-plate tubes of the sixth of an inch in diameter, and two inches long; but they diminished considerably the size of the flame, and rendered it more liable to go out by motion; and the following experiments appear to show, that if the diameter of the upper orifice of the chimney be not very large, it is scarcely possible that any explosion produced by the flame can reach it.

I threw into the safe-lantern with the common

chimney, a mixture of fifteen parts of air, and one of fire-damp: the flame was immediately greatly enlarged, and the flame of the wick seemed to be lost in the larger flame of the fire-damp. I placed a lighted taper above the orifice of the chimney; it was immediately extinguished, but without the slightest previous increase of its flame, and even the wick instantly lost its fire by being plunged into the chimney.

I introduced a lighted taper into a close vessel containing fifteen parts of air, and one of gas from the distillation of coal, suffered it to burn out in the vessel, and then analysed the gas. After the carbonic acid was separated, it appeared by the test of nitrous gas to contain nearly one-third of its original quantity of oxygen; but detonation with a mixture of equal parts of hydrogen and oxygen proved that it contained no sensible quantity of carburetted hydrogen gas.

It is evident then, that when in the safe-lantern, the air gradually becomes contaminated with fire-damp, this fire damp will be consumed in the body of the lantern; and that the air passing through the chimney cannot contain any inflammable mixture.

I made a direct experiment on this point. I gradually threw an explosive mixture of fire-damp and air into the safe-lantern from a bladder furnished with a tube which opened by a large aperture above the flame; the flame became enlarged, and by a rapid jet of gas, I produced an explosion in the body of the lantern; there was not even a current of air through the safety tubes at the moment, and the flame did not appear to reach above the lower aperture of the chimney; and the explosion merely threw out from it a gust of foul air.

The second safety-lantern that I have had made is

upon the same principle as the first, except that instead of tubes, *safety canals* are used, which consist of close concentric hollow metallic cylinders of different diameters, and placed together so as to form circular canals of the diameter of from one-twenty-fifth to one-fortieth of an inch, and an inch and seven-tenths long, by which air is admitted in much larger quantities than by small tubes. In this arrangement there is so free a circulation of air, that the chimney likewise may be furnished with safety-canals.

I have had lamps made for this kind of lantern which stand on the outside, and which may be supplied with oil and cotton without any necessity of opening the lantern; and in this case the chimney is soldered to the top, and the lamp is screwed to the bottom, and the wick rises above the air-canals.

I have likewise had glass lamps with a single wick, and argand lamps made on the same principle, the chimney being of glass covered with a metallic top containing the safety-canals, and the air entering close to the flame through the circular canals.

The third kind of safe lamp or lantern, and which is by far the most simple, is a close lamp or lantern into which the air is admitted, and from which it passes through apertures covered with *brass wire gauze* of $\frac{1}{200}$ of an inch in thickness, the apertures of which should not be more than $\frac{1}{120}$ of an inch; this stops explosions as well as long tubes or canals, and yet admits of a free draught of air.

Having succeeded in the construction of safe lanterns and lamps, equally portable with common lanterns and lamps, which afforded sufficient light, and which bore motion perfectly well, I submitted them individually to practical tests, by throwing into them explosive atmospheres of fire-damp and air. By the natural action of the flame drawing air through the air-canals, from the explosive atmosphere, the light was uniformly extinguished; and when an explosive mixture was forcibly pressed into the body of the lamp, the explosion was always stopped by the safety apertures, which may be said figuratively to act as a sort of *chemical fire sieves* in separating flame from air. But I was not contented with these trials, and I submitted the safe canals, tubes, and wire-gauze fire-sieves, to much more severe tests: I made them the medium of communication between a large glass vessel filled with the strongest explosive mixture of carburetted hydrogen and air, and a bladder two-thirds or one-half full of the same mixture, both insulated from the atmosphere. By means of wires passing near the stop-cock of the glass vessel, I fired the explosive mixture in it by the discharge of a Leyden jar. The bladder always expanded at the moment the explosion was made; a contraction as rapidly took place; and a lambent flame played round the mouths of the safety apertures, open in the glass vessel; but the mixture in the bladder did not explode: and by pressing some of it into the glass vessel so as to make it replace the foul air, and subjecting it to the electric spark, repeated explosions were produced, proving the perfect security of the safety apertures; even when acted on by a much more powerful explosion than could possibly occur from the introduction of air from the mines.

These experiments held good whatever were the proportions of the explosive mixture and whatever was the size of the glass vessel, (no one was ever used containing more than a quart) provided as many as twelve metallic tubes were used of one-seventh of an inch in diameter, and two and a half inches long; or provided the circular metallic canals, were $\frac{1}{25}$ of an inch in diameter, one and $\frac{1}{7}$ of an inch deep, and at least two inches in circumference; or provided the wire-gauze had apertures of only $\frac{1}{120}$ of an inch. When twelve metallic tubes were employed as the medium of communication, $\frac{1}{7}$ of an inch in diameter and an inch long, the explosion was communicated by them into the bladder. Four glass tubes of the $\frac{1}{10}$ of an inch in diameter and two inches long, did not communicate the explosion; but *one* of this diameter and length produced the effect. The explosion was stopped by a single tube $\frac{1}{28}$ of an inch in diameter,* when it was three inches long, but not when it was two inches long.

The explosion was stopped by the metallic gauze of $\frac{1}{120}$ when it was placed between the exploding vessel and the bladder, though it did not present a surface of more than half a square inch, and the explosive mixture in the bladder in passing through it to supply the vacuum produced in the glass vessel burnt on the surface exposed to the glass vessel for some seconds, producing a murmuring noise.

A circular canal $\frac{1}{25}$ of an inch in diameter, an inch and half in circumference, and one and $\frac{7}{10}$ of an inch deep, communicated explosion, but four concentric canals of the same depth and diameter, and of which the smallest was two inches in circumference, and separated from each other only by their sides, which were of brass, and

* These results appear at first view contradictory to those mentioned page 27. But it must be kept in view that the first set of experiments were made in tubes open in the air, and the last in tubes exposed to the whole force of air explosion, and connected only with close vessels filled with explosive mixtures.

about $\frac{17}{40}$ of an inch in thickness, did not suffer the explosion to act through them.

It would appear then, that the smaller the circumference of the canal, that is, the nearer it approaches to a tube, the greater must be its depth, or the less its diameter to render it safe.

I did not perceive any difference in these experiments, when the metals of the apertures were warmed by repeated explosions; it is probable, however, that considerable elevation of temperature would increase the power of the aperture to pass the explosion; but the difference between the temperature of flame, and that marked on our common mercurial scale, is so great that the addition of a few degrees of heat probably does not diminish perceptibly the cooling power of a metallic surface, with regard to flame.

By diminishing the diameter of the air canals their power of passing the explosion is so much diminished that their depth and circumference may be brought extremely low. I found that flame would not pass through a canal of the $\frac{1}{70}$ of an inch in diameter, when it was one-fourth of an inch deep, and forming a cylinder of only one-fourth of an inch in circumference; and a number of apertures of $\frac{1}{100}$ of an inch are safe when their depth is equal to their diameter. It is evident from these facts, that metallic doors, or joinings in lamps may be easily made safe by causing them to project upon and fit closely to parallel metallic surfaces.

Longitudinal air canals of metal may, I find, be employed with the same security as circular canals; and a few pieces of tin-plate soldered together with wires to regulate the diameter of the canal, answer the purpose of the feeder or safe chimney as well as drawn cylinders of brass.

A candle will burn in a lantern or glass tube made safe with metallic gauze, as well as in the open air; I conceive, however, that oil lamps, in which the wick will always stand at the same height, will be preferred.

But the principle applies to every kind of light, and its entire safety is demonstrated.

When the fire-damp is so mixed with the external atmosphere as to render it explosive, the light in the safe-lantern or lamp will be extinguished, and warning will be given to the miners to withdraw from, and to ventilate that part of the mine.

If it be necessary to be in a part of the mine where the fire-damp is explosive, for the purpose of clearing the workings, taking away pillars of coal, or other objects, the workmen may be lighted by a fire made of charcoal, which burns without flame, or by the steel mill, though this does not afford such entire security from danger as the charcoal fire.

It is probable, that when explosions occur from the sparks from the steel mill, the mixture of the fire-damp is in the proportion required to consume all the oxygen of the air, for it is only in about this proportion that explosive mixtures can be fired by electrical sparks from a common machine.

As the wick may be moved without communication between the air in the safe-lantern or lamp and the atmosphere, there is no danger in trimming or feeding them; but they should be lighted in a part of the mine where there is no fire-damp, and by a person charged with the care of the lights; and by these inventions, used with such simple precautions, there is every reason to believe a number of lives will be saved, and much misery prevented. Where candles are employed in the open air in the mines, life is extinguished by the explosion; with the safe-lantern or safe-lamp the light is only put out, and no other inconvenience will occur.

It does not appear, by what I have learnt from the miners, that breathing an atmosphere containing a certain mixture of fire-damp near or even at the explosive point, is attended with any bad consequence. I ascertained that a bird lived in a mixture of equal parts of fire-damp and air; but he soon began to show symptoms of suffering. I found a slight head-ache produced by breathing for a few minutes an explosive mixture of fire-damp and air: and if merely the health of the miners be considered, the fire-damp ought always to be kept far below the point of its explosive mixture.

Miners are sometimes found alive in a mine after explosion has taken place: this is easily explained, when it is considered that the inflammation is almost always limited to a particular spot, and that it mixes the residual air with much common air; and supposing 1 of fire-damp to 13 of air to be exploded, there will still remain nearly one-third of the original quantity of oxygen in the residual gas: and in some experiments, made 16 years ago, I found that an animal lived, though with suffering, for a short time in a gas containing 100 parts of azote, 14 parts of carbonic acid, and 7 parts of oxygen.

EXPLANATION OF THE PLATE.

Plate I.

Fig. 1.—represents the safe-lantern, with its air-feeder and chimney furnished with safety metallic canals. It contains about a quart of air. The sides are of horn or glass, made air-tight by putty or cement. A. is the lamp through which the circular air-feeding canals pass: they are three concentric hollow cylinders, distant from each other $\frac{1}{26}$ of an inch: the smallest is $2\frac{1}{2}$ inches in circumference; their depth is 2 inches. B. is the chimney, containing 4 such canals, the smallest 2 inches in circumference: above it is a hollow cylinder, with a cap to prevent dust from passing into the chimney. C. is the hole for admitting oil. D. is a long canal, containing a wire, by which the wick is moved or trimmed. E. is the tube forming a connection between the reservoir of oil, and the chamber that supplies the wick with oil. F. is the rim round the bottom of the lantern, to enable it to bear motion.

Fig. 2.—is the lamp of fig. 1. of its natural size; the references to the letters are the same.

Fig. 3.—is a common chimney, which may be used in the lantern; but the safety-chimney doubles the security.

Fig. 4.—exhibits the safety concentric canals, or fire-sieves, which, if $\frac{1}{22}$ of an inch in diameter, must not be less than 2 inches in exterior circumference, and 1·7 of an inch high.

Fig. 5.—exhibits the longitudinal safe-canals, or fire-sieves.

Fig. 6.—exhibits a safe-lamp, having a glass chimney covered with tin-plate, and the safety-apertures in a

PLATE 1.

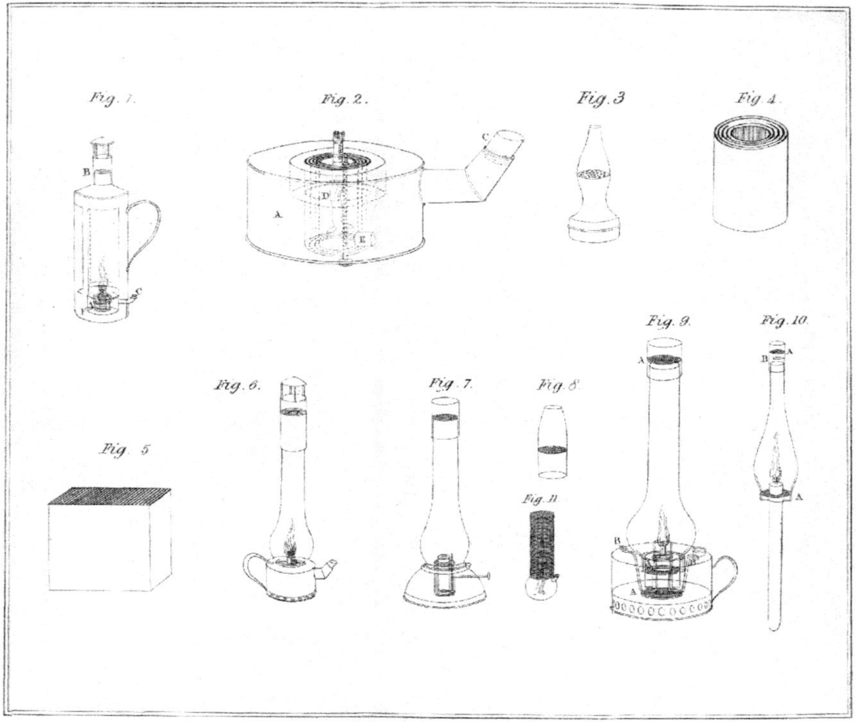
Fig. 1.
B
C
Fig. 2.
A
C
D
E
Fig. 3
Fig. 4.
Fig. 5
Fig. 6.
Fig. 7.
Fig. 8.
Fig. 11.
Fig. 9.
A
B
Fig. 10.
A
B

cylinder, with a covering above: the lower part is the same as in the lantern.

Fig. 7.—an argand lamp, of similar construction with safe air canals without the flame, and metallic gauze apertures within.

Fig. 8.—a tin-plate chimney top for the lamp, made safe by metallic gauze.

Fig. 9.—a metallic gauze safe-lamp. A A A. Screens of metallic gauze, or *flame-sieves.* B B. Wires for trimming the wick.

Fig. 10.—a glass tube furnished with flame-sieves, in which a common candle may be burnt. A A. The flame-sieves. B. A little plate of metal, to prevent the upper flame-sieve from being acted on by the current of hot air.

The lamps burn brighter the higher the chimney.

From my experiments it appears, that a mere narrow throat and opening to the metallic part of the chimney is sufficient to prevent explosions from passing through the lamp, supposing them to be possible; but with safety-canals or metallic gauze in the chimney the security is absolute.

The circular canals, and the apertures covered with metallic gauze, are so much superior to tubes in practical applications, that I have no doubt of their being generally used; I have therefore given no sketch of the first safe-lantern I had constructed with tubes; but substituting tubes for canals, it is exactly the same as that represented, fig. 1.

APPENDIX.

1. In the beginning of my inquiries, I had another close lantern made, which may be called the fire-valve lantern. In this, the candle, or lamp burns, with its

full quantity of air, admitted from an aperture below, till the air begins to be mixed with fire-damp; when, as the fire-damp increases the flame, a thermometrical spring at the top of the lantern, made of brass and steel, riveted together, and in a curved form, expands, moves a valve in the chimney, diminishes the circulation of air, and extinguishes the flame; but I did not pursue this invention, after I had discovered the properties of the fire-damp, on which the safe-lantern is founded.

2. The safety of close lamps, or lanterns, may probably be likewise secured by sieves made of asbestus, or possibly even of hair or silk, placed over the air apertures; but metallic gauze will be necessary above in the chimney. I have little doubt but that windows of fine metallic gauze may be used for giving light in lanterns, with perfect security; perhaps for the chimney it may be worth while to have fine silver plated wire-gauze made.

3. The expansive powers of the fire-damp, during its explosion, are so small, as to render no precautions, with respect to the thickness of the glass or horn in the lamps or lanterns, necessary.

---

2. *An Account of a Method for giving Light in explosive Mixtures of Fire Damp in Coal Mines, by consuming the Fire Damp.**

I have already had the honour of communicating to the Royal Society an account of a Safe Lamp, which

* [From Phil. Trans. for 1816. Read before the Royal Society, January 11, 1816.]

becomes extinguished when introduced in very explosive mixtures of fire-damp.

In this communication I shall describe a light that will burn in any explosive mixture of fire-damp, and the light of which arises from the combustion of the fire-damp itself.

The invention consists in covering or surrounding a flame of a lamp or candle, by a wire sieve; the coarsest that I have tried, with perfect safety contained 625 apertures in a square inch, and the wire was $\frac{1}{70}$ of an inch in thickness; the finest 6400 apertures in a square inch: and the wire was $\frac{1}{250}$ of an inch in diameter.

When a lighted lamp or candle screwed into a ring, soldered to a cylinder of wire gauze, having no apertures, except those of the gauze, or safe apertures, is introduced into the most explosive mixture of carburetted hydrogen and air, the cylinder becomes filled with a bright flame, and this flame continues to burn as long as the mixture is explosive. When the carburetted hydrogen is to the air as 1 to 12, the flame of the wick appears within the flame of the fire-damp; when the proportion is as high as 1 to 7, the flame of the wick disappears.

When the thickest wires are used in the gauze, it becomes strongly red-hot, particularly at the top; but yet no explosion takes place. The flame is brighter the larger the apertures of the gauze; and the cylinder of 625 apertures to the square inch, gives a brilliant light in a mixture of 1 part of gas from the distillation of coal and 7 parts of air. The lower part of the flame is green, the middle purple, and the upper part blue.

I have tried cylinders of 6400 apertures to the square inch, in mixtures of oxygen and carburetted hydrogen, and even in mixtures of oxygen and hydrogen, and though the wire became intensely red-hot, yet explosions never took place; the combustion was entirely limited to the interior of the lamp.

In all these experiments, there was a noise like that produced by the burning of hydrogen gas in open tubes.

These extraordinary and unexpected results, lead to many enquiries respecting the nature and communication of flame; but my object at present is only to point out their application to the use of the collier.

All that he requires to ensure security, are small wire cages*to surround his candle or his lamp, which may be made for a few pence, and of which various modifications may be adopted, and the applications of this discovery will not only preserve him from the fire-damp, but enable him to apply it to use, and to destroy it at the same time that it gives him a useful light.

3. *On the Combustion of Explosive Mixtures confined by Wire Gauze; with some Observations on Flame.*†

I have pursued my inquiries respecting the limits of the size of the apertures and of the wire in the metallic gauze, which I have applied to secure the coal miners from the explosions of fire-damp. Gauze made of brass wire, $\frac{1}{50}$ of an inch in thickness, and containing only 10 apertures to the inch, or 100 apertures in the square inch, employed in the usual way as a guard of flame, did not communicate explosion in a mixture of 1 part of coal-gas and 12 parts of air, as long as it was

* Fig 11, Plate I. represents this contrivance.

† [From the Phil. Trans. for 1816. Read before the Royal Society, January 25, 1816.]

cool; but as soon as the top became hot, an explosion took place.

A quick lateral motion likewise enabled it to communicate explosion.

Gauze, made of the same wire, containing 14 apertures to the inch, or 196 to the square inch, did not communicate explosion till it became strongly red-hot, when it was no longer safe in explosive mixtures of coal-gas; but no motion that could be given to it, by shaking it in a close jar, produced explosion.

Iron wire gauze of $\frac{1}{40}$, and containing 240 apertures in the square inch, was safe in explosive mixtures of coal-gas, till it became strongly red-hot at the top.

Iron wire gauze of $\frac{1}{50}$, and of 24 apertures to the inch, or 576 to the square inch, appeared safe under all circumstances in explosive mixtures of coal-gas. I kept up a continual flame in a cylinder of this kind, 8 inches high and 2 inches in diameter, for a quarter of an hour, varying the proportions of coal-gas and air as far as was compatible with their inflammation; the top of the cylinder, for some minutes, was strongly red-hot; but though the mixed gas was passed rapidly through it by pressure from a gasometer and a pair of double bellows, so as to make it a species of blast furnace, yet no explosion took place.

I mentioned, in my last communication to the Society, that a flame confined in a cylinder of very fine wire gauze, did not explode a mixture of oxygen and hydrogen, but that the gases burnt in it with great vivacity. I have repeated this experiment in nearly a pint of the most explosive mixtures of the two gases; they burnt violently within the cylinder; but, though the upper part became nearly white-hot, yet no explosion was communicated, and it was necessary to withdraw the cylinder, to prevent the brass wire from being melted.

These results are best explained by considering the nature of the flame of combustible bodies—which, in all cases, must be considered as the combustion of an *explosive mixture* of inflammable gas, or vapour and air; for it cannot be regarded as a mere combustion at the surface of contact of the inflammable matter: and the fact is proved, by holding a taper or a piece of burning phosphorus within a large flame, made by the combustion of alcohol, the flame of the candle or of the phosphorus will appear in the centre of the other flame, proving that there is oxygen, even in its interior part.

The heat communicated by flame, must depend upon its mass; this is shown by the fact, that the top of a slender cylinder of wire-gauze hardly ever becomes dull red in the experiment on an explosive mixture; whilst in a larger cylinder, made of the same material, the central part of the top soon becomes bright red. A large quantity of cold air thrown upon a small flame, lowers its heat beyond the explosive point; and, in extinguishing a flame, by blowing upon it, the effect is probably principally produced by this cause.

If a piece of wire-gauze sieve is held over a flame of a lamp or of coal-gas, it prevents the flame from passing it, and the phenomenon is precisely similar to that exhibited by the wire-gauze cylinders; the air passing through is found very hot, for it will convert paper into charcoal; and it is an explosive mixture, for it will inflame, if a lighted taper be presented to it; but it is cooled below the explosive point, by passing through wires even red-hot, and by being mixed with a considerable quantity of air comparatively cold. The real temperature of visible flame, is perhaps as high as any we are acquainted with. Mr. Tennant was in the habit of showing an experiment, which demonstrates the intensity of its heat. He used to fuse a small filament of platinum in the flame of a common candle; and it is proved by many facts, that a stream of air may be made to render a metallic body white-hot, yet not be itself luminous.

A considerable mass of heated metal is required to inflame even coal-gas, or the contact of the same mixture with an extensive heated surface. An iron wire of $\frac{1}{20}$ of an inch, and 8 inches long, red-hot, when held perpendicularly in a stream of coal-gas, did not inflame it, nor did a short wire of $\frac{1}{6}$ of an inch produce the effect held horizontally; but wire of the same size, when six inches of it were red-hot, and when it was held perpendicularly in a bottle, containing an explosive mixture, so that heat was successively communicated to portions of the gas, produced its explosion.

A certain degree of mechanical force which rapidly throws portions of cold explosive mixture upon flame, prevents explosions at the point of contact; thus on pressing an explosive mixture of coal gas from a syringe, or a gum elastic bottle, it burns only at some distance from the aperture from which it is disengaged.

Taking all these circumstances into account, there appears no difficulty in explaining the combustion of explosive mixture within and not without the cylinders; for a current is established from below upwards, and the hottest part of the cylinder is where the results of combustion, the water, carbonic acid, or azote, which are not inflammable, pass out. The gas which enters is not sufficiently heated on the outside of the wire, to be exploded, and as the gases are no where confined, there can be no mechanical force pressing currents of flame towards the same point.

It will be needless to enter into further illustrations of the theoretical part of the subject: and I shall conclude this paper by stating what I am sure will be gratifying to the Society, that the cylinder lamps have been tried in two of the most dangerous mines near Newcastle, with perfect success; and from the communications I have had from the collieries, there is every reason to believe that they will be immediately adopted in all the mines in that neighbourhood, where there is any danger from fire-damp.

4. *Notice of some New Views and Experiments respecting Flame.**

When a wire-gauze safe-lamp is made to burn in a very explosive mixture of coal gas and air, the light is feeble, and of a pale colour; whereas the flame of a current of coal gas burnt in the atmosphere, as is well known by the phenomena of the gas-lights, is extremely brilliant. In a paper read before the Royal Society, I have endeavoured to show, that in all cases flame is a continued combustion of explosive mixtures; it becomes, therefore, a problem of some interest, "Why the combustion of explosive mixtures, under different circumstances, should produce such different appearances?" A very acute philosopher, who himself started the subject in conversation, suggested the idea, that in the combustion of explosive mixtures within the lamp, carbonic oxide might be formed; and that the light might be deficient, from the deficiency of the quantity of oxygen necessary to produce carbonic acid. On submitting this idea to the test of experiment, it was dis-

* [From the Journal of Science and the Arts, vol. ii. p. 124.]

covered to be unfounded; for, by the combustion in the wire-gauze lamp, carbonic acid was produced in quantities as great as could have been expected from the quantity of oxygen consumed; and on adding oxygen to a mixture in quantities more than sufficient to burn the whole of the gas, the character of the light still continued the same.

In reflecting on the circumstances of the two species of combustion, I was led to imagine that the cause of the superiority of the light of the *stream* of coal gas might be owing to the *decomposition* of a part of the gas towards the interior of the flame where the air was in smallest quantity, and the deposition of solid charcoal, which, first by its *ignition*, and afterwards by its *combustion*, increased in a high degree the intensity of the light; and a few experiments soon convinced me that this was the true solution of the problem.

I held a piece of wire-gauze, of about 900 apertures to the square inch, over a stream of coal-gas issuing from a small pipe, and inflamed the gas above the wire-gauze, which was almost in contact with the orifice of the pipe; when it burned with its usual bright light. On raising the wire-gauze so as to cause the gas to be mixed with more air before it inflamed, the light became feebler; and at a certain distance the flame assumed the precise character of that of an explosive mixture burning within the lamp; but though the light was so feeble in this last case, the heat was greater than when the light was much more vivid, and a piece of wire of platinum held in this feeble blue flame became instantly white hot.

On reversing the experiment by inflaming a stream of coal gas, and passing a piece of wire-gauze gradually from the summit of the flame to the orifice of the pipe, the result was still more instructive; for it was found that the apex of the flame intercepted by the wire-gauze afforded no solid charcoal; but in passing it downwards, solid charcoal was given off in considerable quantities, and prevented from burning by the cooling agency of the wire-gauze; and at the bottom of the flame, where the gas burnt blue in its immediate contact with the atmosphere, charcoal ceased to be deposited in visible quantities.

This principle of the increase of the brilliancy and density of flame by the production and ignition of solid matter, appears to admit of many applications.

1st. It explains readily the appearances of the different parts of the flames of burning bodies, and of flame urged by the blow-pipe; the point of the inner blue flame, where the heat is greatest, is the point where the whole of the charcoal is burnt in its gaseous combinations without previous deposition.

2dly. It explains the intensity of the light of those *flames* in which *fixed* solid matter is produced in combustion, such as that of the flame of phosphorus* and of zinc in oxygen, &c., and of potassium in chlorine; and the feebleness of the light of those flames in which gaseous and volatile matter alone is produced, such as those of hydrogen and sulphur in oxygen, phosphorus in chlorine, &c.

3rdly. It offers means of increasing the light of certain burning substances, by placing in their flames even incombustible substances. Thus the intensity of the light of burning sulphur, carbonic oxide, &c. is won-

* Since this paper has been written I have found that phosphoric acid volatilizes slowly at a strong red heat, but under moderate pressure it bears a white heat, and in a flame so intense as that of phosphorus, the elastic force must produce the effect of compression.

derfully increased by throwing into them oxide of zinc, or by placing in them very fine amianthus or metallic gauze.

4thly. It leads to deductions respecting the chemical nature of bodies and various phenomena of their decomposition. Thus ether burns with a flame which seems to indicate the presence of olefiant gas in that substance. Alcohol burns with a flame similar to that of a mixture of carbonic oxide and hydrogen; so that the first is probably a binary compound of olefiant gas and water, and the second of carbonic oxide and hydrogen.

When cuprane or protochloride of copper is introduced into the flame of a candle or lamp, it affords a peculiar dense and brilliant red light, tinged with green and blue towards the edges, which seems to depend upon the chlorine being separated from the copper by the hydrogen, and the ignition and combustion of the solid copper and charcoal.

Similar explanations may be given of the phenomena presented by the action of other combinations of chlorine on flame; and it is probable, in many of those cases when the colour of flame is changed by the introduction of incombustible compounds, that the effect depends upon the production and subsequent ignition or combustion of inflammable matter from them. Thus the rose-coloured light given to flame by the compounds of strontium and calcium,* and the yellow colour given by those of barium, and the green by those of boron, may depend upon a temporary production of these bases by the inflammable matter of the flame.

* A similar effect I find is produced by the compounds of the new fixed alkali, lithia, which serves to distinguish its compounds from those of potassa and soda.—1818.

Whenever a flame is remarkably brilliant and dense, it may be always concluded that some solid matter is produced in it: on the contrary, when a flame is extremely feeble and transparent, it may be inferred that no solid matter is formed. Thus none of the volatile combinations of sulphur burn with a flame in the slightest degree opaque; and, consequently, there is no reason, from the phenomena of its flame, to suspect the existence of any fixed basis in sulphur.

5thly. These views will probably offer illustrations of electrical light. The voltaic arc of flame from the great battery, differs in colour and intensity according as the substances employed in the circuit are different; and is infinitely more brilliant and dense with charcoal than with any other substance. May not this depend upon particles of the substances separated by the electrical attractions? and the particles of charcoal being the lightest amongst solid bodies, (as their elementary proportional number shews,) and the least coherent, would be separated in the largest quantities.

6thly. The heat of flames may be actually diminished by increasing their light, (at least the heat communicable to other matter,) and *vice versa.* The flame from combustion which produces the most intense heat amongst those I have examined, is that of a mixture of oxygen and hydrogen in slight excess, compressed in a blowpipe apparatus, and inflamed from a tube having a very small aperture.* This flame is hardly visible in bright

* John George Children, Esq., first proposed to me this application of the blow-pipe by compression, immediately after I discovered that the explosion from oxygen and hydrogen might be arrested by very small apertures, and I first tried the experiment with a fine glass capillary tube. The flame was *not visible* at the end of this tube, being overpowered by the brilliant star of the glass ignited at the aperture. The results produced by the action of this flame on solid bodies, were so similar

day-light, yet it instantly fuses very refractory bodies; and the light from solid matters ignited in it, is so vivid as to be painful to the eye.

July 1816.

5. *Some New Researches on Flame.**

I have described in three papers which the Royal Society have honoured with a place in their Transactions, a number of experiments on combustion, which show that the explosion of gaseous mixtures can be prevented or arrested by various cooling influences, and which led me to discover a tissue permeable to light and air, but impermeable to flame, on which I founded the invention of the wire-gauze safe-lamp now generally used in all collieries in which inflammable air prevails, for the preservation of the lives and persons of the miners. In a short notice published in the third number of the Journal of Science and the Arts, edited at the Royal Institution, I have given an account of some new results on flame, which show that the intensity of the light of flames depends principally upon the production and ignition of solid matter in combustion, and that the heat and light in this process are in a great measure independent phenomena. Since this notice has been printed, I have made a number of researches on flame: and as they appear to me to throw some new lights on this important subject, and to lead to some practical views connected with the useful arts, I shall without any farther apology, present them to the Royal Society.

to those published by Mr. Hare some years ago, that I did not think them worthy of particular notice.

* [From the Phil. Trans. for 1817. Read before the Royal Society, Jan. 16, 1817.]

That greater distinctness may exist in the details, I hall treat of my subjects under four heads. In the first I shall discuss the effects of rarefaction, by partly removing the pressure of the atmosphere upon flame and explosion. In the second, I shall consider the effects of heat in combustion. In the third, I shall examine the effect of the mixture of gaseous substances not concerned in combustion upon flame and explosion. In the fourth I shall offer some general views upon flame, and point out certain practical and theoretical applications of the results.

I. ON THE EFECT OF RAREFACTION BY PARTLY REMOVING THE PRESSURE OF THE ATMOSPHERE UPON FLAME AND EXPLOSION.

The earlier experimenters upon the Boylean vacuum observed that flame ceased in highly rarefied air: but the degree of rarefaction necessary for this effect, has been differently stated. Amongst late experimenters, M. de Grotthus has examined this subject. He has asserted that a mixture of oxygen and hydrogen ceases to be explosive by the electrical spark when rarefied sixteen times, and that a mixture of chlorine and hydrogen cannot be exploded when rarefied only six times, and he generalises by supposing that rarefaction, whether produced by removing pressure or by heat, has the same effect.

I shall not begin by discussing the experiments of this ingenious author. My own results and conclusions are very different from his; and the cause of this difference will I think be obvious in the course of these inquiries. I shall proceed in stating the observations which guided my researches.

When hydrogen gas slowly produced from a proper

mixture was inflamed at a fine orifice of a glass tube, as in the experiment called the philosophical candle, so as to make a jet of flame of $\frac{1}{6}$ of an inch in height, and introduced under the receiver of an air pump containing from 200 to 300 cubical inches of air, the flame enlarged as the receiver became exhausted, and when the gage indicated a pressure between 4 and 5 times less than that of the atmosphere, was at its maximum of size; it then gradually diminished below, but burned above till the pressure was between 7 and 8 times less, when it became extinguished.

To ascertain whether the effect depended upon the deficiency of oxygen, I used a larger jet with the same apparatus, when the flame to my surprise burned longer, and when the atmosphere was rarefied ten times, and this in repeated trials. When the larger jet was used, the point of the glass tube became white-hot, and continued red-hot till the flame was extinguished. It immediately occurred to me, that the heat communicated to the gas by this tube, was the cause that the com bustion continued longer in the last trials when the larger flame was used; and the following experiments confirmed the conclusion. A piece of wire of platinum was coiled round the top of the tube, so as to reach into and above the flame. The jet of gas of $\frac{1}{6}$ of an inch in height was lighted and the exhaustion made; the wire of platinum soon became white-hot in the centre of the flame, and a small point of wire near the top fused: it continued white-hot till the pressure was 6 times less, when it was 10 times it continued red-hot at the upper part, and, as long as it was dull red, the gas though extinguished below, continued to burn in contact with the hot wire, and the combustion did not cease until the pressure was reduced 13 times.

It appears from this result, that the flame of hydrogen is extinguished in rarefied atmospheres, only when the heat it produces is insufficient to keep up the combustion, which appears to be when it is incapable of communicating visible ignition to metal, and as this is the temperature required for the inflammation of hydrogen at common pressures, it appears that its *combustibility* is neither diminished nor increased by rarefaction from the removal of pressure.

According to this view with respect to hydrogen, it should follow that amongst other combustible bodies, those which require least heat for their combustion, ought to burn in more rarefied air than those that require more heat, and those that produce much heat in their combustion ought to burn, other circumstances being the same, in more rarefied air than those that produce little heat: and every experiment I have made confirms these conclusions. Thus olefiant gas, which approaches nearly to hydrogen in the heat produced by its combustion, and which does not require a much higher temperature for its inflammation, when its flame was made by a jet of gas from a bladder connected with a small tube furnished with a wire of platinum, under the same circumstances as hydrogen, ceased to burn when the pressure was diminished between 10 and 11 times: and the flames of alcohol and of the wax taper which require a greater consumption of heat for the volatilization and decomposition of their combustible matter, were extinguished when the pressure was 5 or 6 times less without the wire of platinum, and 7 or 8 time less when the wire was kept in the flame. Light carburetted hydrogen, which produces, as will be seen hereafter, less heat in combustion than any of the common combustible gases, except carbonic oxide, and which requires a higher temperature for its inflammation than any other, had its flame extinguished, even though the tube was furnished with the wire when the pressure was below $\frac{1}{4}$.

The flame of carbonic oxide which, though it produces little heat in combustion, is as inflammable as hydrogen, burned when the wire was used, the pressure being $\frac{1}{6}$.

The flame of sulphuretted hydrogen, the heat of which is in some measure carried off by the sulphur produced by its decomposition during its combustion in rare air, when burned in the same apparatus as the olefiant and other gases, was extinguished when the pressure was $\frac{1}{7}$.

Sulphur, which requires a lower temperature for its combustion than any common inflammable substance, except phosphorus, burned with a very feeble blue flame in air rarefied fifteen times, and at this pressure the flame heated a wire of platinum, to dull redness, nor was it extinguished till the pressure was reduced to $\frac{1}{20}$.*

Phosphorus, as has been shown by M. Van Marum, burns in an atmosphere rarefied 60 times; and I found that phosphuretted hydrogen produced a flash of light when admitted into the best vacuum that could be made, by an excellent pump of Nairn's construction.

* The temperature of the atmosphere diminishes in a certain ratio with its height, which must be attended to in the conclusions respecting combustion in the upper regions of the atmosphere, and the elevation must be somewhat lower than in arithmetical progression, the pressure decreasing in geometrical progression.

There is, however, every reason to believe, that the taper would be extinguished at a height of between 9 and 10 miles, hydrogen between 12 and 13, and sulphur between 15 and 16.

The mixture of chlorine and hydrogen inflames at a much lower temperature than that of hydrogen and oxygen, and produces a considerable degree of heat in combustion; it was therefore probable that it would bear a greater degree of rarefaction, without having its power of exploding destroyed; and this I found in many trials is actually the case, contrary to the assertion of M. de Grotthus. Oxygen and hydrogen in the proportion to form water, will not explode by the electrical spark when rarefied eighteen times, but hydrogen and chlorine in the proportion to form muriatic acid gas, gave a distinct flash of light under the same circumstances, and they combined with visible inflammation when the spark was passed through them, the exhaustion being to $\frac{1}{24}$.

The experiment on the flame of hydrogen with the wire of platinum, and which holds good with the flames of the other gases, shows, that by preserving heat in rarefied air, or giving heat to a mixture, inflammation may be continued when, under common circumstances, it would be extinguished. This I found was the case in other instances, when the heat was differently communicated: thus, when camphor was burned in a glass tube, so as to make the upper part of the tube red-hot, the inflammation continued when the rarefaction was 9 times, whereas it would only continue in air rarefied 6 times, when it was burned in a thick metallic tube which could not be considerably heated by it.

By bringing a little naphtha in contact with a red-hot iron, it produced a faint lambent flame, when there remained in the receiver only $\frac{1}{30}$ of the original quantity of air, though without foreign heat its flame was extinguished when the quantity was $\frac{1}{6}$.

I rarefied a mixture of oxygen and hydrogen by the

air pump to about eighteen times, when it could not be inflamed by the electric spark. I then heated strongly the upper part of the tube till the glass began to soften, and passed the spark, when a feeble flash was observed not reaching far into the tube, the heated gases only appearing to enter into inflammation. This last experiment requires considerable care. If the exhaustion is much greater, or if the heat is raised very slowly,* it does not succeed; and if the heat is raised so high as to make the glass luminous, the flash of light, which is extremely feeble, is not visible: it is difficult to procure the proper degree of exhaustion, and to give the exact degree of heat; I have, however, succeeded three times in obtaining the results, and in one instance it was witnessed by Mr. Brande.

To elucidate the inquiry still farther, I made a series of experiments on the heat produced by some of the inflammable gases in combustion. In comparing the heat communicated to wires of platinum by flames of the same size, it was evident that hydrogen and olefiant gas in oxygen, and hydrogen in chlorine, produced a much greater intensity of heat in combustion, than the other gaseous substances I have named burned in oxygen: but no regular scale could be formed from observations of this kind. I endeavoured to gain some approximations on the subject by burning equal quantities of different gases under the same circumstances, and applying the heat to an apparatus by which it could be measured. For this purpose a mercurial gas-holder was furnished with a system of stop-cocks, terminating in a strong tube of platinum having a minute aperture. Above this was fixed a copper cup filled with olive-oil, in which a thermometer was placed. The oil was

* The reason will be obvious from what is stated in page 61.

heated to 212° to prevent any differences in the communication of heat by the condensation of aqueous vapour; the pressure was the same for the different gases, and they were consumed as nearly as possible in the same time, and the flame applied to the same point of the copper cup, the bottom of which was wiped after each experiment.

The results were as follows:

The flame from olefiant gas raised the therm. to	270°
hydrogen	238
sulphuretted hydrogen	232
coal gas	236
gaseous oxide of carbon	218

The quantities of oxygen consumed (that absorbed by the hydrogen being taken as 1) would be, supposing the combustion perfect, for the olefiant gas 6, for the sulphuretted hydrogen 3, for the carbonic oxide 1. The coal-gas contained only a very small proportion of olefiant gas; supposing it to be pure carburetted hydrogen, it would have consumed 4 of oxygen. Taking the elevations of temperature, and the quantities of oxygen consumed as the data, the ratios of the heat produced by the combustion of the different gases, would be for hydrogen 26, for olefiant gas 9·66, for sulphuretted hydrogen 6·66, for carburetted hydrogen 6, for carbonic oxide 6.*

It will be useless to reason upon this ratio as exact, for charcoal was deposited both from the olefiant gas and coal-gas during the experiment, and much sulphur was deposited from the sulphuretted hydrogen; and there is great reason to believe, that the capacities of

* These results may be compared with Mr. Dalton's new System of Chemical Philosophy; they agree in showing that hydrogen produces more heat in combustion than any of its compounds.

fluids for heat increase with their temperature. It confirms, however, the general conclusions, and proves that hydrogen stands at the head of the scale, and gaseous oxide of carbon at the bottom. It might at first view be imagined that, according to this scale, the flame of carbonic oxide ought to be extinguished by rarefaction, at the same degree as that of carburetted hydrogen; but it must be remembered, as I have mentioned in another place, that carbonic oxide is a much more combustible gas. Carbonic oxide inflames in the atmosphere when brought into contact with an iron wire heated to dull redness, whereas carburetted hydrogen is not inflammable by a similar wire, unless it is heated to whiteness so as to burn with sparks.

II. ON THE EFFECTS OF RAREFACTION BY HEAT AND EXPLOSION.

The results detailed in the preceding section are indirectly opposed to the opinion of M. de Grotthus, that rarefaction by heat destroys the combustibility of gaseous mixtures. Before I made any direct experiments on this subject, I endeavoured to ascertain the degree of expansion which can be communicated to elastic fluids by the strongest heat that can be applied to glass vessels. For this purpose I introduced into a graduated curved glass tube some fusible metal. I heated the fusible metal and the portion of the tube containing the air included by it, under boiling water for some time. I then placed the apparatus in a charcoal fire, and very gradually raised the temperature till the fusible metal appeared luminous when viewed in the shade. At this time the air had expanded so as to occupy 2·25 parts in the tube, it being 1 at the temperature of boiling-water. Another experiment was made in a thicker glass tube, and the heat was raised until the tube began to run together; but though this heat appeared cherry-red, the expansion was not to more than 2·5, and a part of this might perhaps have been apparent only, owing to the collapsing of the glass tube before it actually melted. It may be supposed that the oxidation of the fusible metal may have had some effect in making the expansion appear less; but in the first experiment the air was gradually brought back to its original temperature of boiling water, when the absorption was scarcely sensible. If M. Gay Lussac's conclusions be taken as the groundwork of calculation, and it be supposed that air expands equally for equal increments of temperature, it would appear that the temperature of air capable of rendering glass luminous must be 1035° Fahrenheit.*

M. de Grotthus describes an experiment in which atmospheric air and hydrogen, expanded to four times their bulk over mercury by heat, would not inflame by the electric spark. It is evident, that in this experiment a large quantity of steam or of mercurial vapour must have been present, which, like other inexplosive elastic fluids, prevents combustion when mixed in certain quantities with explosive mixtures: but though he seems aware that his gases were not dry, yet he draws his general conclusions, that expansion by heat destroys the explosive powers of gases, principally from this inconclusive experiment.

I introduced into a small graduated tube over well

* The mode of ascertaining temperatures as high as the point of fusion of glass by the expansion of air, seems more unexceptionable than any other. It gives for the point of visible ignition nearly the same degree as that deduced by Newton from the times of the cooling of ignited metal in the atmosphere.

boiled mercury, a mixture of two parts of hydrogen and one of oxygen, and heated the tube by a large spirit lamp till the volume of the gas was increased from 1 to 2·5. I then, by means of a blow-pipe and another spirit-lamp, made the upper part of the tube red-hot, when an explosion instantly took place.

I introduced into a bladder a mixture of oxygen and hydrogen, and connected this bladder with a thick glass tube of about $\frac{1}{6}$ of an inch in diameter and three feet long, curved so that it could be gradually heated in a charcoal furnace; two spirit lamps were placed under the tube, where it entered the charcoal fire, and the mixture was very slowly pressed through: an explosion took place before the tube was red-hot.

This experiment shows that expansion by heat, instead of diminishing the combustibility of gases, on the contrary, enables them to explode, apparently, at a lower temperature, which seems perfectly reasonable, as a part of the heat communicated by any ignited body must be lost in gradually raising the temperature. I made several other experiments, which establish the same conclusions. A mixture of common air and hydrogen was introduced into a small copper tube, having a stopper not quite tight; the copper tube was placed in a charcoal fire; before it became visibly red, an explosion took place, and the stopper was driven out.

I made various experiments on explosions, by passing mixtures of hydrogen and oxygen through heated tubes; in the beginning of one of these trials, in which the heat was much below redness, steam appeared to be formed without any combustion. This led me to expose mixtures of oxygen and hydrogen in tubes, in which they were confined by fluid fusible metal to heat: and I found, that by carefully applying a heat between the boiling point of mercury, which is not sufficient for the effect, and a heat approaching to the greatest heat that can be given without making glass luminous in darkness, the combination was effected without any violence, and without any light: and commencing with 212°, the volume of steam formed at the point of combination appeared exactly equal to that of the original gases. So that the first effect in experiments of this kind is an expansion, afterwards a contraction, and then the restoration of the primitive volume.

If, when this change is going on, the heat be quickly raised to redness, an explosion takes place; but with small quantities of gas the change is completed in less than a minute.

It is probable, that the slow combination without combustion, already long ago observed with respect to hydrogen and chlorine, oxygen and metals, will happen at certain temperatures with most substances that unite by heat. On trying charcoal, I found, that at a temperature, which appeared to be a little above the boiling point of quicksilver, it converted oxygen pretty rapidly into carbonic acid, without any luminous appearance, and at a dull red heat, the elements of olefiant gas combined in a similar manner with oxygen, slowly and without explosion.

The effect of the slow combination of oxygen and hydrogen is not connected with their rarefaction by heat: for I found that it took place when the gases were confined in a tube by fusible metal, rendered solid at its upper surface; and certainly as rapidly, and without any appearance of light.

M. de Grotthus has stated, that, if a glowing coal be brought into contact with a mixture of oxygen and hydrogen, it only rarefies them, but does not explode them; but this depends upon the degree of heat communicated by the coal: if it is red in daylight, and free from ashes, it uniformly explodes the mixture; if its redness is barely visible in shade, it will not explode them, but cause their slow combination; and the general phenomenon is wholly unconnected with rarefaction, as is shown by the following circumstance:—When the heat is greatest, and before the invisible combination is completed, if an iron wire, heated to whiteness, be placed upon the coal within the vessel, the mixture instantly explodes.

Light carburetted hydrogen, or pure fire-damp, as has been shown, requires a very strong heat for its inflammation; it therefore offered a good substance for an experiment on the effect of high degrees of rarefaction by heat on combustion. I mixed together one part of this gas and eight parts of air, and introduced them into a bladder furnished with a capillary tube. I heated this tube till it began to melt, and then slowly passed the mixture through it into the flame of a spirit lamp, when it took fire and burned with its own peculiar explosive light beyond the flame of the lamp, and when withdrawn, though the aperture was quite white-hot, it continued to burn vividly.

That the compression in one part of an explosive mixture produced by the sudden expansion of another part by heat, or the electric spark, is not the cause of combination, as has been supposed by Dr. Higgins, M. Berthollet, and others, appears to be evident from what has been stated, and it is rendered still more so by the following facts. A mixture of hydro-phosphoric gas (bi-phosphuretted hydrogen gas) and oxygen, which explode at a heat a little above that of boiling-water, was confined by mercury, and very gradually heated on a sand-bath; when the temperature of the mercury was 242°, the mixture exploded.

A similar mixture was placed in a receiver communicating with a condensing syringe, and condensed over mercury till it occupied only $\frac{1}{5}$ of its original volume. No explosion took place, and no chemical change had occurred, for when its volume was restored, it was instantly exploded by the spirit lamp.

It would appear, then, that *the heat* given out by the compression of gases is the real cause of the combustion which it produces, and that at certain elevations of temperature, whether in rarefied or compressed atmospheres, explosion or combustion occurs, i. e. bodies combine with the production of heat and light.

III. ON THE EFFECTS OF THE MIXTURE OF DIFFERENT GASES IN EXPLOSION AND COMBUSTION.

In my first paper on the fire-damp of coal mines, I have mentioned that carbonic acid gas has a greater power of destroying the explosive power of mixtures of fire-damp and air than azote, and I have ventured to suppose the cause to be its greater density and capacity for heat, in consequence of which it might exert a greater cooling agency, and prevent the temperature of the mixture from being raised to that degree necessary for combustion. I have lately made a series of experiments with the view of determining how far this idea is correct, and for the purpose of ascertaining the general phenomena of the effects of the mixture of gaseous substances upon explosion and combustion.

I took given volumes of a mixture of two parts of hydrogen and one part of oxygen by measure, and diluting them with various quantities of different elastic fluids, I ascertained at what degree of dilution the power of inflammation by a strong spark from a Leyden phial was destroyed. I found that for one of the mixture inflammation was prevented by

Of Hydrogen, about	8
Oxygen	9
Nitrous oxide	11
Carburetted hydrogen	1
Sulphuretted hydrogen	2
Olefiant gas	$\frac{1}{2}$
Muriatic acid gas	2
Silicated fluoric acid gas	$\frac{5}{6}$

Inflammation took place when the mixtures contained

Of Hydrogen	6
Oxygen	7
Nitrous oxide	10
Carburetted hydrogen	$\frac{3}{4}$
Olefiant gas	$\frac{1}{3}$
Sulphuretted hydrogen	$1\frac{1}{2}$
Muriatic acid gas	$1\frac{1}{2}$
Fluoric acid gas	$\frac{3}{4}$

I hope to be able to repeat these experiments with more precision at no distant time; the results are not sufficiently exact to lay the foundation for any calculations on the relative cooling powers of equal volumes of the gases, but they show sufficiently, if the conclusions of MM. De la Roche and Berard be correct, that other causes, besides density and capacity for heat, interfere with the phenomena. Thus nitrous oxide, which is nearly $\frac{1}{3}$ denser than oxygen, and which, according to De la Roche and Berard, has a greater capacity for heat in the ratio of 1·3503 to ·9765 in volume, has lower powers of preventing explosion; and hydrogen, which is 15 times lighter than oxygen, and which in equal volumes has a smaller capacity for heat, certainly has a higher power of preventing explosion; and olefiant gas exceeds all other gaseous substances in a much higher ratio than could have been expected from its density and capacity. The olefiant gas I used was recently made, and might have contained some vapour of ether, and the nitrous oxide was mixed with some azote, but these slight causes could not have interfered with the results to any considerable extent.

Mr. Leslie, in his elaborate and ingenious researches on heat, has observed the high powers of hydrogen of abstracting heat from solid bodies, as compared with that of common air and oxygen. I made a few experiments on the comparison of the powers of hydrogen, in this respect, with those of carburetted hydrogen, azote, oxygen, olefiant gas, nitrous oxide, chlorine, and carbonic acid gas. The same thermometer raised to the same temperature, 160°, was exposed to equal volumes (21 cubic inches) of olefiant gas, coal-gas, carbonic acid gas, chlorine, nitrous oxide gas, hydrogen, oxygen, azote, and air, at equal temperatures, 52° Fahrenheit.

The times required for cooling to 106° were for

	′ ″
Air	2
Hydrogen	45
Olefiant gas	1·15
Coal-gas	55
Azote	1·30
Oxygen	1·47
*Nitrous oxide	2·30· 2·53
*Carbonic acid gas	2·45
Chlorine	3·6

It appears from these experiments, that the powers of

* These two last results were observed by Mr. Faraday of the Royal Institution, (from whom I receive much useful assistance in most of my experiments), when I was absent from the Laboratory.

elastic fluids to abstract or conduct away heat from solid surfaces, is in some inverse ratio to their density, and that there is something in the constitution of the light gases, which enables them to carry off heat from solid surfaces in a different manner from that in which they would abstract it in gaseous mixtures, depending probably upon the mobility of their parts.* The heating of gaseous media by the contact of fluid or solid bodies, as has been shown by Count Rumford, depends principally upon the change of place of their particles; and it is evident from the results stated in the beginning of this section, that these particles have different powers of abstracting heat analogous to the different powers of solids and fluids. Where an elastic fluid exerts a cooling influence on a cold surface, the effect must depend principally upon the rapidity with which its particles change their places: but where the cooling particles are mixed throughout a mass with other gaseous particles, their effect must principally depend upon the power they possess of rapidly abstracting heat from the contiguous particles; and this will depend probably upon two causes, the simple abstracting power by which they become quickly heated, and their capacity for heat which is great in proportion as their temperatures are less raised by this abstraction.

Whatever be the cause of the different cooling powers of the different elastic fluids in preventing inflammation, very simple experiments show that they operate uniformly with respect to the different species of combustion, and that those explosive mixtures, or in-

* Those particles which are lightest must be conceived most capable of changing place, and would therefore cool solid surfaces most rapidly: in the cooling of gaseous mixtures, the mobility of the particles can be of little consequence.

flammable bodies, which require least heat for their combustion, require larger quantities of the different gases to prevent the effect, and *vice versa;* thus one of chlorine and one of hydrogen still inflame when mixed with eighteen times their bulk of oxygen, whereas a mixture of carburetted hydrogen and oxygen in the proper proportions for combinations, one and two, have their inflammation prevented by less than three times their volume of oxygen.

A wax taper was instantly extinguished in air mixed with $\frac{1}{10}$ of silicated fluoric acid gas, and in air mixed with $\frac{1}{6}$ of muriatic acid gas; but the flame of hydrogen burned readily in those mixtures, and in mixtures in which the flame of hydrogen was extinguished, the flame of sulphur burned.

There is a very simple experiment which demonstrates in an elegant manner this general principle. Into a long bottle with a narrow neck introduce a lighted taper, and let it burn till it is extinguished: carefully stop the bottle, and introduce another lighted taper, it will be extinguished before it reaches the bottom of the neck: then introduce a small tube containing zinc and diluted sulphuric acid, and at the aperture of which the hydrogen is inflamed; the hydrogen will be found to burn in whatever part of the bottle the tube is placed: after the hydrogen is extinguished, introduce lighted sulphur; this will burn for some time, and after its extinction, phosphorus will be as luminous as in the air, and, if heated in the bottle, will produce a pale yellow flame of considerable density.

In cases when the heat required for chemical union is very small, as in the instance of hydrogen and chlorine, a mixture which prevents inflammation will not prevent combination, i. e. the gases will combine without any flash. This I witnessed in mixing two volumes of carburetted hydrogen with one of chlorine and hydrogen: muriatic acid was formed throughout the mixture, and heat produced, as was evident from the expansion when the spark passed, and the rapid contraction afterwards, but the heat was so quickly carried off by the quantity of carburetted hydrogen that no flash was visible.

In the case of phosphorus, which is combustible at the lowest temperature of the atmosphere, no known admixture of elastic fluid prevents the luminous appearance; but this seems to depend upon the light being limited to the solid particles of phosphoric acid formed; whereas to produce flame, a certain mass of elastic fluid must be luminous; and there is every reason to believe, that when phosphuretted hydrogen explodes in very rare air, it is only the phosphorus which is consumed. Any other substance that produces solid matter in combustion would probably be luminous in air as rare, or in mixtures as diluted, as phosphorus, provided the heat was elevated sufficiently for its combustion. I have found that this is actually the case with respect to zinc. I threw some zinc filings into an ignited iron crucible fixed on the stand of an air pump under a receiver, and exhausted until only $\frac{1}{60}$ of the original quantity of air remained. When I judged that the red-hot crucible must be full of the vapour of zinc, I admitted about $\frac{1}{60}$ more of air, when a bright flash of light took place in and above the crucible, similar to that which is produced by admitting air to the vapour of phosphorus in vacuo.

The cooling power of mixtures of elastic fluids in preventing combustion must increase with their condensation, and diminish with their rarefaction; at the same time, the quantity of matter entering into combustion in given spaces, is relatively increased and diminished. The experiments on flame in rarefied atmospherical air, show that the quantity of heat produced in combustion is very slowly diminished by rarefaction, the diminution of the cooling power of the azote being apparently in a higher ratio than the diminution of the heating powers of the burning bodies. I endeavoured to ascertain what would be the effect of condensation on flame in atmospheric air, aud whether the cooling power of the azote would increase in a lower ratio, as might be expected, than the heat produced by the increase of the quantity of matter, entering into combustion, but I found considerable difficulties in making the experiments with precision. I ascertained, however, that both the light and heat of the flames of the taper, of sulphur and hydrogen, were increased by acting on them by air condensed four times; but not more than they would have been by an addition of $\frac{1}{5}$ of oxygen.

I condensed air nearly five times, and ignited iron wire to whiteness in it by the voltaic apparatus, but the combustion took place with very little more brightness than in the common atmosphere, and would not continue as in oxygen, nor did charcoal burn much more brightly in this compressed air than in common air. I intend to repeat these experiments, if possible, with higher condensing powers: they show sufficiently, that, (for certain limits at least) as rarefaction does not diminish considerably the heat of flame in atmospherical air, so neither does condensation considerably increase it; a circumstance of great importance in the constitution of our atmosphere, which at all the heights or depths at which man can exist, still preserves the same relations to combustion.

It may be concluded from the general law, that at high temperatures, gases not concerned in combustion will have less powers of preventing that operation, and likewise that steam and vapours, which require a considerable heat for their formation, will have less effect in preventing combustion, particularly of those bodies requiring low temperatures, than gases at the common heat of the atmosphere.

I have made some experiments on the effects of steam, and their results were conformable to these views. I found that a very large quantity of steam was necessary to prevent sulphur from burning. Oxygen and hydrogen exploded by the electric spark, when mixed with five times their volume of steam; and even a mixture of air and carburetted hydrogen gas, the least explosive of all mixtures, required a third of steam to prevent its explosion,—whereas $\frac{1}{5}$ of azote produced the effect. These trials were made over mercury: heat was applied to water above the mercury, and 37·5 for 100 parts was regarded as the correction for the expansion of the gases.

It is probable that, with certain heated mixtures of gases, where the non-supporting or non-inflammable elastic fluids are in great quantities, combination with oxygen will take place, as in the instance mentioned, page 69, of hydrogen and chlorine, without any light —for the temperature produced will not be sufficient to render elastic media luminous; and there are no combustions, except those of the compounds of phosphorus and the metals in which solid matters are the result of combinations with oxygen. I have shown in the paper referred to in the introduction, that the light of common flames depends almost entirely upon the deposition, ignition and combustion of solid charcoal; but to produce this deposition from gaseous substances, demands a high temperature. Phosphorus, which rises in vapour at common temperatures, and the vapour of which combines with oxygen at those temperatures, as I have mentioned before, is always luminous; for each particle of acid formed must, there is every reason to believe, be white-hot; but so few of these particles exist in a given space, that they scarcely raise the temperature of a solid body exposed to them, though, as in the rapid combustion of phosphorus, where immense numbers are existing in a small space, they produce a most intense heat.

In all cases the quantity of heat communicated by combustion, will be in proportion to the quantity of burning matter coming in contact with the body to be heated. Thus the blow-pipe and currents of air operate. In the atmosphere, the effect is impeded by the mixture of azote, though still it is very great: with pure oxygen compression produces an immense effect; and with currents of oxygen and hydrogen, there is every reason to believe that solid matters are made to attain the temperature of the flame. This temperature, however, evidently presents the limit to experiments of this kind, for bodies exposed to flame can never be hotter than flame itself; whereas, in the voltaic apparatus, there seems to be no limit to the heat, except the volatilization of the conductors.

The temperatures of flames are probably very different. Where, in chemical changes, there is no change of volume, as, in the instance of the mutual action of chlorine and hydrogen, prussic gas (cyanogen) and oxygen, approximations to their temperatures may be gained from the expansion in explosion.

I have made some experiments of this kind, by detonating the gases by the electrical spark in a curved tube containing mercury or water; and I judged of the expansion, from the quantity of fluid thrown out of the tube; the resistance opposed by mercury, and its great cooling powers, rendered the results very unsatisfactory in the cases in which it was used; but with water, cyanogen, and oxygen being employed, they were more conclusive. Cyanogen and oxygen, in the proportion of 1 to 2, detonated in a tube of about two-fifths of an inch in diameter, displaced a quantity of water, which demonstrated an expansion of 15 times their original bulk. This would indicate a temperature of above 5000° of Fahrenheit, and the real temperature is probably much higher; for heat must be lost by communication to the tube and the water. The heat of the gaseous carbon in combustion in this gas, appears more intense than that of hydrogen; for I found a filament of platinum was fused by a flame of cyanogen in the air, which was not fused by a similar flame of hydrogen.

IV. SOME GENERAL OBSERVATIONS, AND PRACTICAL INFERENCES.

The knowledge of the cooling power of elastic media, in preventing the explosion of the fire-damp, led me to those practical researches, which terminated in the discovery of the wire-gauze safe lamp; and the general investigations of the relation and extent of these powers, serves to elucidate the operation of wire-gauze and other tissues or systems of apertures permeable to light and air, in intercepting flame, and confirms the views I originally gave of the phenomenon.

Flame is gaseous matter, heated so highly as to be luminous, and that to a degree of temperature beyond the white heat of solid bodies, as is shown by the circumstance that air, not luminous, will communicate this degree of heat.* When an attempt is made to pass flame through a very fine mesh of wire-gauze at the common temperature, the gauze cools each portion of the elastic matter that passes through it, so as to reduce its temperature below that degree at which it is luminous, and the diminution of temperature must be proportional to the smallness of the mesh, and the mass of the metal. The power of a metallic or other tissue to prevent explosion, will depend upon the heat required to produce the combustion, as compared with that acquired by the tissue; and the flame of the most inflammable substances, and of those that produce most heat in combustion, will pass through a metallic tissue that will interrupt the flame of less inflammable substances, or those that produce little heat in combustion. Or the tissue being the same, and impermeable to all flames at common temperatures, the flames of the most combustible substances, and of those which produce most heat, will most readily pass through it when it is heated, and each will pass through it at a different degree of temperature. In short, all the circumstances which apply to the effect of cooling mixtures upon flame, will apply to cooling perforated surfaces. Thus, the flame of phosphuretted hydrogen at common temperatures, will pass through a tissue sufficiently large not to be immediately choked up by the phosphoric acid formed, and the phosphorus deposited.† A tissue of 100 apertures to

* This is proved by the simple experiment of holding a fine wire of platinum about the one-twentieth of an inch from the exterior of the middle of the flame of a spirit lamp, and concealing the flame by an opaque body, the wire will become white-hot in a space where there is no visible light.

† If a tissue containing above 700 apertures to the square inch be held over the flame of phosphuretted hydrogen, it does not transmit the flame till it is sufficiently heated to enable the phosphorus to pass through it

the square inch, made of wire of $\frac{1}{60}$, will at common temperatures intercept the flame of a spirit lamp, but not that of hydrogen ; and when strongly heated, it will no longer arrest the flame of the spirit lamp. A tissue which will not interrupt the flame of hydrogen, when red-hot, will still intercept that of olefiant gas, and a heated tissue which would communicate explosion from a mixture of olefiant gas and air, will stop an explosion from a mixture of fire-damp, or carburetted hydrogen.

The ratio of the combustibility of the different gaseous matters are likewise, to a cetain extent, as the masses of heated matter required to inflame them.* Thus an iron wire of $\frac{1}{40}$ of an inch heated cherry-red, will not inflame olefiant gas, but it will inflame hydrogen gas; and a wire of $\frac{1}{8}$, heated to the same degree, will inflame olefiant gas; but a wire of $\frac{1}{500}$ must be heated to whiteness to inflame hydrogen, though at a low red heat it will inflame bi-phosphuretted gas; but wire of $\frac{1}{40}$, heated even to whiteness, will not inflame mixtures of fire-damp.

These circumstances will explain, why a mesh of wire so much finer, is required to prevent the explosion from hydrogen and oxygen from passing, and why so coarse a texture and wire is sufficient to prevent the explosion of the fire-damp, fortunately the least combustible of the known inflammable gases.

The general doctrine of the operation of wire-gauze cannot be better elucidated than in its effects upon the

in vapour. Phosphuretted hydrogen is decomposed in flame, and acts exactly like phosphorus.

* It appeared to me, in these experiments, that the worst conducting and best radiating substances required to be heated higher for equal masses to produce the same effect upon the gases; thus, red-hot charcoal had evidently less power of inflammation than red-hot iron.

flame of sulphur. When wire-gauze of 600 or 700 apertures to the square inch is held over the flame, fumes of condensed sulphur immediately come through it, and the flame is intercepted; the fumes continue for some instants, but as the heat increases they diminish; and at the moment they disappear, which is long before the gauze becomes red-hot, the flame passes; the temperature at which sulphur burns being that at which it is gaseous.

Another very simple illustration of the truth of this view is offered in the effect of the cooling agency of metallic surfaces upon very small flames. Let the smallest possible flame be made by a single thread of cotton immersed in oil, and burning immediately upon the surface of the oil; it will be found to be about $\frac{1}{30}$ of an inch in diameter. Let a fine iron wire of $\frac{1}{180}$ be made into a circle of $\frac{1}{10}$ of an inch in diameter, and brought over the flame. Though at such a distance, it will instantly extinguish the flame, if it be *cold:* but if it be held above the flame, so as to be slightly heated, the flame may be passed through it without being extinguished. That the effect depends entirely upon the power of the metal to abstract the heat of flame, is shown by bringing a glass capillary ring of *the same* diameter and size over the flame; this being a much worse conductor of heat, will not extinguish it even when *cold.* If its size, however, be made greater, and its circumference smaller, it will act like the metallic wire, and require to be heated to prevent it from extinguishing the flame.*

* Let a small globe of metal of 1-20th of an inch in diameter, made by fusing the end of a wire, be brought near a flame of 1-30th in diameter, it will extinguish it when cold, at the distance of its own diameter; let it be heated, and the distance will diminish at which it produces the extinction; and at a white heat it does not extinguish it by actual contact, though at a dull red heat it immediately produces the effect.

Suppose a flame, divided by the wire-gauze into smaller flames, each flame must be extinguished in passing its aperture till that aperture has attained a temperature sufficient to produce the permanent combustion of the explosive mixture.

A flame of sulphur may be made much smaller than that of hydrogen, that of hydrogen smaller than that of a wick fed with oil, and that of a wick fed with oil smaller than that of carburetted hydrogen; and a ring of cool wire, which instantly extinguishes the flame of carburetted hydrogen, only slightly diminishes the size of a flame of sulphur of the same dimensions.

Where rapid currents of explosive mixtures are made to act upon wire-gauze, it is of course much more rapidly heated; and therefore the same mesh which arrests the flames of explosive mixtures at rest, will suffer them to pass when in rapid motion; but by *increasing* the cooling surface by diminishing the size, or increasing the depth of the aperture, all *flame*, however rapid their motion, may be arrested. Precisely the same law applies to explosions acting in close vessels: very minute apertures, when they are only few in number, will permit explosions to pass, which are arrested by much larger apertures when they fill a whole surface. A small aperture was drilled at the bottom of a wire-gauze lamp in the cylindrical ring which confines the wire-gauze; this, though less than $\frac{1}{18}$ of an inch in diameter, passed the flame, and fired the external atmosphere, in consequence of the whole force of the explosion of the thin stratum of the mixture included within the cylinder driving the flame through the aperture; though, had the whole ring been composed of such apertures, separated by wires, it would have been perfectly safe.

Nothing can demonstrate more decidedly than these

simple facts and observations, that the interruption of flame by solid tissues, permeable to light and air, depends upon no recondite or mysterious cause, but to their cooling powers, simply considered as such.

When a light included in a cage of wire-gauze is introduced into an explosive atmosphere of fire-damp at rest, the maximum of heat is soon obtained, the radiating power of the wire, and the cooling effect of the atmosphere, more efficient from the mixture of inflammable air, prevents it from ever arriving at a temperature equal to that of dull redness. In rapid currents of explosive mixtures of fire-damp, which heat common gauze to a higher temperature, twilled gauze, in which the radiating surface is considerably greater, and the circulation of air less, preserves an equal temperature. Indeed, the heat communicated to the wire by combustion of the fire-damp in wire-gauze lamps, is completely in the power of the manufacturer; for by diminishing the apertures, and increasing the mass of metal, or the radiating surface, it may be diminished to any extent.

I have lately had lamps made of thick twilled gauze of wires of $\frac{1}{40}$, sixteen to the warp, and thirty to the weft, which being rivetted to the screw, cannot be displaced; from its flexibility it cannot be broken, and from its strength cannot be crushed, except by a very strong blow.

Even in the common lamps the flexibility of the material has been found of great importance, and I could quote one instance of a dreadful accident having been prevented, which must have happened had any other material than wire-gauze been employed in the construction of the lamp: and how little difficulty has occurred in the practical application of the invention, is shown by the circumstance, that it has been now for ten months in the hands of hundreds of common miners in the most dangerous mines in Britain, during which time not a single accident has occurred where it has been employed, whilst in other mines, much less dangerous, where it has not yet been adopted, some lives have been lost, and many persons burned.*

The facts stated in Section II. explain why so much more heat is obtained from fuel when it is burnt quickly; and they show that in all cases the temperature of the acting bodies should be kept as high as possible, not only because the general increment of heat is greater, but likewise, because those combinations are prevented which at lower temperatures take place without any considerable production of heat: thus, in the Argand lamp, the Liverpool lamp, and in the best fire-places, the increase of effect does not depend merely upon the rapid current of air, but likewise upon the heat preserved by the arrangements of the materials of the chimney, and communicated to the matters entering into inflammation.

These facts likewise explain the methods by which temperature may be increased, and the limit to certain methods. Currents of flame, as it was stated in the last section, can never raise the heat of bodies exposed to them, higher than a certain degree, their own temperature; but by compression, there can be no doubt, the heat of flames from pure supporters and combustible

* Plates of different forms of this lamp are annexed. (Pl. II.) They are applicable to all purposes in which explosions or inflammations are to be guarded against, whether from fire-damp, or carburetted hydrogen, coal gas, vapours of spirits, or of ether. And by the introduction of glass cylinders within the wire-gauze cylinder above the flame, the wick may be made very large, and it burns on the principle of the Liverpool lamp.

matter may be greatly increased, probably in the ratio of their compression. In the blow-pipe of oxygen and hydrogen, the maximum of temperature is close to the aperture from which the gases are disengaged, i. e. where their density is greatest. Probably a degree of temperature far beyond any that has been yet attained may be produced by throwing the flame from compressed oxygen and hydrogen into the voltaic arc, and thus combining the two most powerful agents for increasing temperature.

The circumstances mentioned in this Paper, combined with those noticed in the Paper on flame printed in Mr. Brande's Journal of Science and the Arts, explain the nature of the light of flames and their form. When in flames pure gaseous matter is burnt, the light is extremely feeble: the density of a common flame is proportional to the quantity of solid charcoal first deposited and afterwards burnt. The form of the flame is conical, because the greatest heat is in the centre of the explosive mixture. In looking steadfastly at flame, the part where the combustible matter is volatilized is seen, and it appears dark, contrasted with the part in which it begins to burn, that is, where it is so mixed with air as to become explosive. The heat diminishes towards the top of the flame, because in this part the quantity of oxygen is least. When the wick increases to a considerable size from collecting charcoal, it cools the flame by radiation, and prevents a proper quantity of air from mixing with its central part; in consequence, the charcoal thrown off from the top of the flame is only red-hot, and the greater part of it escapes unconsumed.

The intensity of the light of flames in the atmosphere is increased by condensation, and diminished by rarefaction, apparently in a higher ratio than their heat, more particles capable of emitting light exist in the denser atmospheres, and yet most of these particles in becoming capable of emitting light, absorb heat; which could not be the case in the condensation of a pure supporting medium.

The facts stated in Section I. show that the luminous appearances of shooting stars and meteors cannot be owing to any inflammation of *elastic* fluids, but must depend upon the ignition of solid bodies. Dr. Halley calculated the height of a meteor at ninety miles, and the great American meteor which threw down showers of stones, was estimated at seventeen miles high. The velocity of motion of these bodies must in all cases be immensely great, and the heat produced by the compression of the most rarefied air from the velocity of motion must be probably sufficient to ignite the mass; and all the phenomena may be explained, if *falling stars* be supposed to be small solid bodies moving round the earth in very eccentric orbits, which become ignited only when they pass with immense velocity through the upper regions of the atmosphere, and if the *meteoric bodies* which throw down stones with explosions be supposed to be similar bodies which contain either combustible or elastic matter.

*Some new Experiments and Observations on the Combustion of Gaseous Mixtures, &c.**

In a paper read before the Royal Society at their last two meetings, I have described the phenomena of the

* [From the Phil. Trans. for 1817, read before the Royal Society, January 23, 1817.]

slow combustion of hydrogen and olefiant gas without flame. In the same paper I have shown, that the temperature of flame is infinitely higher than that necessary for the ignition of solid bodies. It appeared to me, therefore, probable, that in certain combinations of gaseous bodies, for instance, those above referred to, when the increase of temperature was not sufficient to render the gaseous matters themselves luminous; yet still it might be adequate to ignite solid matters exposed to them. I had devised several experiments on this subject. I had intended to expose fine wires to oxygen and olefiant gas, and to oxygen and hydrogen during their slow combination under different circumstances, when I was accidentally led to the knowledge of the *fact*, and, at the same time, to the discovery of a new and curious series of phenomena.

I was making experiments on the increase of the limits of the combustibility of gaseous mixtures of coal gas and air by increase of temperature. For this purpose, I introduced a small wire-gauze safe-lamp with some fine wire of platinum fixed above the flame, into a combustible mixture containing the maximum of coal gas, and when the inflammation had taken place in the wire-gauze cylinder, I threw in more coal gas, expecting that the heat acquired by the mixed gas in passing through the wire-gauze would prevent the excess from extinguishing the flame. The flame continued for two or three seconds after the coal-gas was introduced; and when it was extinguished, that part of the wire of platinum which had been hottest remained ignited, and continued so for many minutes, and when it was removed into a dark room, it was evident that there was no flame in the cylinder.

It was immediately obvious that this was the result

which I had hoped to attain by other methods, and that the oxygen and coal-gas in contact with the hot wire combined without flame, and yet produced heat enough to preserve the wire ignited, and to keep up their own combustion. I proved the truth of this conclusion by making a similar mixture, heating a fine wire of platinum and introducing it into the mixture. It immediately became ignited nearly to whiteness, as if it had been itself in actual combustion, and continued glowing for a long while, and when it was extinguished, the inflammability of the mixture was found entirely destroyed.

A temperature much below ignition only was necessary for producing this curious phenomenon, and the wire was repeatedly taken out and cooled in the atmosphere till it ceased to be visibly red; and yet when admitted again, it instantly became red-hot.

The same phenomena were produced with mixtures of olefiant gas and air, carbonic oxide, prussic gas and hydrogen, and in the last case with a rapid production of water; and the degree of heat I found could be regulated by the thickness of the wire. The wire, when of the same thickness, became more ignited in hydrogen than in mixtures of olefiant gas, and more in mixtures of olefiant gas than in those of gaseous oxide of carbon.

When the wire was very fine, about the $\frac{1}{80}$ of an inch in diameter, its heat increased in very combustible mixtures, so as to explode them. The same wire in less combustible mixtures only continued bright red, or dull red, according to the nature of the mixture.

In mixtures not explosive by flame within certain limits, these curious phenomena took place whether the air or the inflammable gas was in excess.

The same circumstances occurred with certain inflammable vapours. I have tried those of ether, alcohol, oil of turpentine and naphtha. There cannot be a better mode of illustrating the fact, than by an experiment on the vapour of ether or of alcohol, which any person may make in a minute. Let a drop of ether be thrown into a cold glass, or a drop of alcohol into a warm one. Let a few coils of wire of platinum of the $\frac{1}{60}$ or $\frac{1}{70}$ of an inch be heated at a hot poker or candle, and let it be brought into the glass; it will in some part of the glass become glowing, almost white hot, and will continue so as long as a sufficient quantity of vapour and of air remain in the glass.

When the experiment on the slow combustion of ether is made in the dark, a pale phosphorescent light is perceived above the wire, which of course is most distinct when the wire ceases to be ignited. This appearance is connected with the formation of a peculiar acrid volatile substance possessed of acid properties.

The chemical changes in general produced by slow combustion appear worthy of investigation. A wire of platinum introduced under the usual circumstances into a mixture of prussic gas (cyanogen), and oxygen in excess became ignited to whiteness, and the yellow vapours of nitrous acid were observed in the mixture. And in a mixture of olefiant gas non-explosive from the excess of inflammable gas, much carbonic oxide was formed.

I have tried to produce these phenomena with various metals; but I have succeeded only with platinum and palladium; with copper, silver, iron, gold, and zinc, the effect is not produced. Platinum and palladium have low conducting powers, and small capacities for heat compared with other metals, and these seem to be the principal causes of their producing, continuing, and rendering sensible these slow combustions.

I have tried some earthy substances which are bad conductors of heat; but their capacities and power of radiating heat appear to interfere. A thin film of carbonaceous matter entirely destroys the igniting power of platinum, and a slight coating of sulphuret deprives palladium of this property, which must principally depend upon their increasing the power of the metals to radiate heat.

Thin laminæ of the metals, if their form admits of a free circulation of air, answer as well as fine wires; and a large surface of platinum may be made red-hot in the vapour of ether, or in a combustible mixture of coal-gas and air.*

I need not dwell upon the connection of these facts respecting slow combustion, with the other facts I have described in the history of flame. Many theoretical views will arise from this connection, and hints for new researches, which I hope to be able to pursue in another communication. I shall now conclude by a practical application. By hanging some coils of fine wire of platinum, or a fine sheet of platinum or palladium above the wick of his lamp, in the wire-gauze cylinder, the coal miner, there is every reason to believe, will be supplied with light in mixtures of fire-damp no longer explosive; and should his flame be extinguished by the quantity of fire-damp, the glow of the metal will continue to guide him, and by placing the lamp in different parts of the gallery, the relative brightness of the wire

* [It has since been found that this metal has the property, even when cold, if its surface is perfectly pure, of effecting the combination of oxygen and hydrogen. Vide Mr. Faraday's ingenious experiments on this subject, Phil. Trans. 1834.]

will show the state of the atmosphere in these parts. Nor can there be any danger with respect to respiration whenever the wire continues ignited, for even this phenomenon ceases when the foul air forms about $\frac{2}{5}$ of the volume of the atmosphere.

I introduced into a wire-gauze safe-lamp a small cage made of fine wire of platinum of the $\frac{1}{70}$ of an inch in thickness, and fixed it by means of a thick wire of platinum about two inches above the wick which was lighted. I placed the whole apparatus in a large receiver, in which, by means of a gas holder, the air could be contaminated to any extent with coal gas. As soon as there was a slight admixture of coal gas, the platinum became ignited; the ignition continued to increase till the flame of the wick was extinguished, and till the whole cylinder became filled with flame; it then diminished. When the quantity of coal-gas was increased so as to extinguish the flame; at the moment of the extinction the cage of platinum became white-hot, and presented a most brilliant light. By increasing the quantity of the coal-gas still farther, the ignition of the platinum became less vivid. When its light was barely sensible, small quantities of air were admitted, its heat speedily increased; and by regulating the admission of coal-gas and air it again became white-hot, and soon after lighted the flame in the cylinder, which as usual, by the addition of more atmospherical air, re-kindled the flame of the wick.

This experiment has been very often repeated, and always with the same results. When the wire for the support of the cage, whether of platinum, silver, or copper, was very thick, it retained sufficient heat to enable the fine platinum wire to re-kindle in a proper mixture half a minute after its light had been entirely destroyed by an atmosphere of pure coal-gas; and by increasing its thickness the period might be made still longer.

The phenomenon of the ignition of the platinum takes place feebly in a mixture consisting of two of air and one of coal-gas, and brilliantly in a mixture consisting of three of air and one of coal-gas: the greater the quantity of heat produced the greater may be the quantity of the coal-gas, so that a large tissue of wire will burn in a more inflammable mixture than single filaments, and a wire made white-hot will burn in a more inflammable mixture than one made red-hot. If a mixture of three parts of air and one of fire-damp be introduced into a bottle, and inflamed at its point of contact with the atmosphere, it will not explode, but will burn like a pure inflammable substance. If a fine wire of platinum coiled at its end be slowly passed through the flame, it will continue ignited in the body of the mixture, and the same gaseous matter will be found to be inflammable and to support combustion.

There is every reason to hope that the same phenomena will occur with the cage of platinum in the fire-damp, as those which have been described in its operation on mixtures of coal-gas. In trying experiments in fire-damp, the greatest care must be taken that no filament or wire of platinum protrudes on the exterior of the lamp, for this would fire externally an explosive mixture. However small the mass of platinum which kindles an explosive mixture in the safe-lamp, the result is the same as when large masses are used; the force of the explosion is directed to, and the flame arrested by, the whole of the perforated tissue.

When a large cage of wire of platinum is introduced

into a very small safe-lamp, even explosive mixtures of fire-damp are burnt without flame; and by placing any cage of platinum in the bottom of the lamp round the wick, the wire is prevented from being smoked. I have sent lamps furnished with this apparatus to be tried in the coal mines of Newcastle and Whitehaven: and I anxiously wait for the accounts of their effects in atmospheres in which no other permanent light can be produced by combustion.

EXPLANATION OF PLATE II.

Representing different Forms of the Miners' Safe-lamp, with the Apparatus for giving light in explosive Mixtures.

a represents the single cylinder of wire-gauze; the foldings *a*, *a*, *a*, must be very well doubled and fastened by wire. If the cylinder be of twilled wire-gauze, the wire should be at least of the thickness of $\frac{1}{40}$ of an inch, and of iron or copper, and 30 in the warp and 16 or 18 in the weft. If of plain wire-gauze, the wire should not be less than $\frac{1}{60}$ of an inch in thickness, and from 28 to 30 both warp and woof.

b represents the second top which fits upon *a*.

c represents a cylinder of brass, in which the wire-gauze is fastened by a screw to prevent it from being separated from the lamp by any blow. *c* is fitted into a female screw, which receives the male screw β of the lamp *f*. *f* is the lamp furnished with its safe trimmer and safe feeder for oil.

A, is the wire-gauze lamp put together with its strong wire supports, which may be three or four, receiving the handle.

J is a small cage made of wire of platinum, of $\frac{1}{70}$ or

PLATE 2

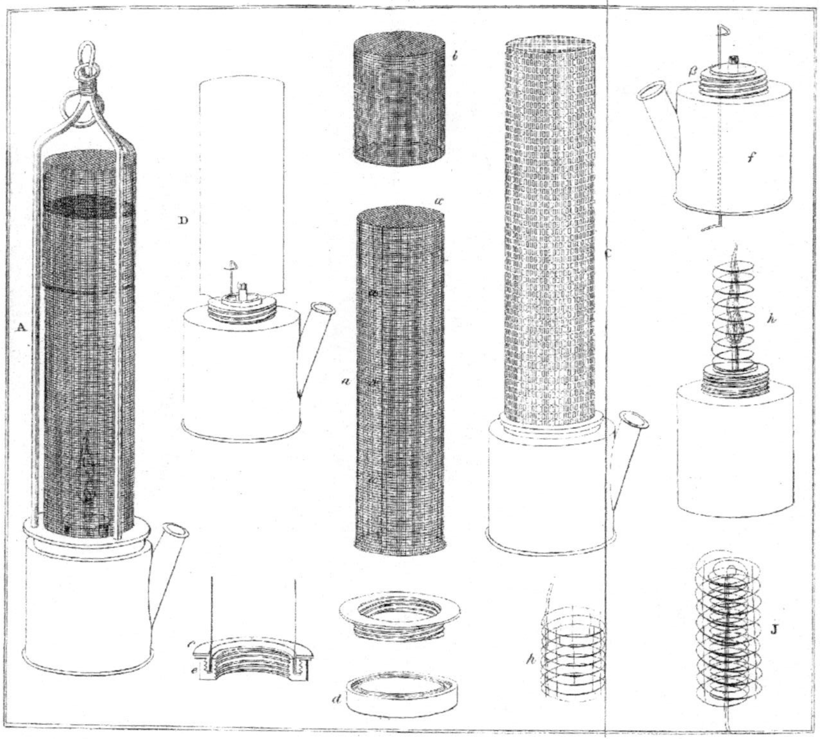

$\frac{1}{80}$ of an inch in thickness, fastened to a wire for raising it above the wick, for giving light in inflammable media, containing too little air to be explosive.

h is a similar cage for placing in the bottom of the lamp, to prevent it from being smoked by the wick.

C is a lamp of which the cylinder is copper of $\frac{1}{40}$ of an inch in thickness, perforated with longitudinal apertures of not more than the $\frac{1}{16}$ of an inch in length, and the $\frac{1}{30}$ in breadth. In proportion as the copper is thicker, the apertures may be increased in size. This form of a lamp may be proper where such an instrument is only to be occasionally used, but for the general purposes of the collier, wire-gauze, from its flexibility, and the ease with which new cylinders are introduced, is much superior.*

D is a lamp fitted with a tin-plate mirror of half the circumference of the cylinder, and reaching as high as the single top, which may be used in strong currents of fire-damp to prevent the heat from rising too high.

All these forms of the wire-gauze lamp are equally safe. In the twilled-gauze lamp less fire-damp is burnt, and the radiating and cooling surface is greater, and it is therefore fitted for very explosive mixtures, or for explosive currents. The wire-gauze lamp with a double cylinder, or with a reflector, answers the same purpose.

The general principle is, that the cylinder should in no case be suffered to be heated above dull redness; and this is always affected by increasing the cooling surfaces, or by diminishing the circulation of the air.

I cannot conclude this notice respecting the safe-

* In the first lamps which I made on this plan, more than twelve months ago, the apertures were circular; but in this case their diameters were required to be very small, as the circular aperture is the most favourable to the transmission of flame.

lamp, without stating, that in the practical application of my views I have received the most enlightened and liberal assistance from the Rev. John Hodgson and Mr. Buddle, who have been the first persons to put my principles to the test of actual experiment in the mines, and to confide their safety to those new resources of chemistry.

III.—SOME EXTRACTS FROM COMMUNICATIONS ON THE APPLICATION OF THE SAFETY LAMP.

The evidence of the use of a practical discovery, is of most value when furnished by practical men. I shall, therefore, annex extracts from some communications on the application of the safety lamp. Those from Newcastle and Whitehaven, will, I am sure, derive importance from the names attached to them. That from Wales is given, as affording an instance in which the combustion of the fire-damp within the lamp was sufficient to destroy it for a considerable time in the workings of a colliery, and it is amusing, from the simplicity of the detail.

1. *Extract from a Letter on the practical Application of the Wire-gauze Safe-lamp, from John Buddle, Esq. to Sir H. Davy.*

Wall's-end Colliery, Newcastle,
1st June, 1816.

"After having introduced your safety-lamp into general use in all the collieries under my direction, where inflammable air prevails, and after using them daily in every variety of explosive mixture for upwards of three months, I feel the highest possible gratification in stating to you that they have answered to my entire satisfaction.

"The safety of the lamps is so easily proved, by taking them into any part of a mine charged with fire-damp, and all the explosive gradations of that dangerous element, are so easily and satisfactorily ascertained by their application, as to strike the minds of the most prejudiced with the strongest conviction of their high utility; and our colliers have adopted them with the greatest eagerness.

"In the practical application of the lamps, scarcely any difficulty has occurred. Those of the ordinary working size, when prepared with common cotton wick and the Greenland whale oil, burn during the collier's *shift*, or day's work of six hours, without requiring to be replenished; and the safety trimmer answers the purpose of cleaning, raising, and lowering the wick completely.

"The only inconvenience experienced, arises from the great quantity of dust, produced in some situations by working the coal, closing up the meshes of the wire-gauze, and obscuring the light; but the workmen very soon removed this inconvenience, by the application of a small brush.

"Our colliers have found it most convenient to hang the stationary lamps from small wooden pedestals; but on observing that where the side of the lamps have been suffered to come in contact with the pedestals, the wood is charred to a considerable depth, by the heat of the lamps; I have thought it right to use small iron pedestals, instead of the wooden ones.

"Besides the facilities afforded by this invention to the working of coal mines, abounding in fire-damp, it has enabled the directors and superintendents to ascertain, with the utmost precision and expedition, both the presence, the quantity, and the correct situation of the gas. Instead of creeping inch by inch with a candle, as is usual, along the galleries of a mine suspected to contain fire-damp, in order to ascertain its presence, we walk firmly on with the safe-lamps, and with the utmost confidence prove the actual state of the mine. By observing attentively the several appearances upon the flame of the lamp, in an examination of this kind, the cause of accidents, which have happened to the most experienced and cautious miners, is completely developed; and this has hitherto been, in a great measure, matter of mere conjecture.

"When the discharge of inflammable air is regular, and the density of the atmosphere continues uniform, the firing point may be judged of, and approached with safety, by a common candle. But when the discharge of inflammable air is irregular, or the atmosphere is in an unsettled state, a degree of uncertainty and danger attends the experiment of ascertaining the state of a mine. With the safe-lamp, however, it is reduced to the utmost certainty, the actual presence and position of the gas is not only ascertained with the greatest precision, but also every alteration of circumstance or position, is distinctly perceived.

"By placing a lamp near the spot where a quantity of inflammable air is issuing, and mixing with the circulating current of atmospherical air to the firing point, it will be seen that very remote causes frequently produce pulsations in the atmosphere of the mine, which occasion the gas to fire at *naked* lights; thus showing clearly the instability of the element with which we have to deal, and the reason why so many explosions have occurred where lights have not approached the place where the gas was lodged within a considerable distance.

"Objections have been made by some who have not had experience of the lamps, to the delicacy of the wire-gauze, under the apprehension that it may be very soon impaired by the flame within the cylinder. Of this, however, I have no reason to complain—as, after three months' constant use, the gauze has not, in the hands of careful workmen, been perceptibly injured by the action of the flame; but the outer top gauze of one or two of Newman's making, has been worn through by the friction of the rivet* on the bottom of the swivel, to which the finger ring is fastened; but this only happened to the lamps used by the *wastemen*, whose business it is to travel daily in the various avenues of the mines, for the purpose of keeping the passage for the current of air free from obstructions: nothing of the kind has happened to the stationary lamps used by the colliers. In short, I do not apprehend that the gauze can be injured by any ordinary cause, without being observed in time sufficient to prevent accidents; and that we have no danger to apprehend, except from the gross negligence of some heedless individual, or an accident of a very unusual description, occurring to injure the gauze.

"I find that I have extended my letter to a greater length than I intended; but I trust, Sir, that you will excuse me for having gone so much into detail, as I feel peculiar satisfaction in dwelling upon a subject which is of the utmost importance, not only to the great cause of humanity, and to the mining interest of this country, but also to the commercial and manufacturing interests of the United Kingdom: for I am convinced that, by

* This rivet is now fixed.—H. D.

the happy invention of the safe-lamp, large proportions of the coal mines of the empire will be rendered available, which otherwise might have remained inaccessible —at least without an invention of similar utility, it could not have been wrought without much loss of the mineral, and risk of life and capital.

"It is not necessary that I should enlarge upon the national advantages which must necessarily result from an invention calculated to prolong our supply of mineral coal, because I think them obvious to every reflecting mind; but I cannot conclude without expressing my highest sentiments of admiration for those talents which have developed the properties and controlled the power of one of the most dangerous elements which human enterprize has hitherto had to encounter."

2. *Extracts from Papers written by John Buddle, Esq., on the Use of the Wire-Gauze Safety-Lamp.*

"Having observed in some of the periodical publications, certain remarks on Sir H. Davy's lamp, which in my mind appear to have originated in motives unconnected with truth and the improvement of science, I feel myself called upon to do an act of justice to the merit of the invention, in a public statement of its great utility and extensive use in the coal mines of this country.

"During the last ten months it has been extensively employed in all the collieries under my inspection; and it gives me the highest pleasure to be able to state, that during that time not the slightest accident by fire has occurred from its use, though several hundreds of lamps are daily employed.

"In the parts of mines where fire-damp prevails, the surveys and inspections are now carried on by the light of the lamp without apprehension of danger from explosion; for experience has shown us, that, with the caution of keeping it in proper repair, it is absolutely safe; and for the truth of this, I appeal to all my professional brethren who have had occasion to use it, without fear of contradiction.

"The colliers never hesitate a moment to take it into any respirable part of a mine, however much it may be charged with fire-damp; for, whenever it appears that the air, either from discharges of gas, or from casual interruptions of the circulating current, becomes explosive, only give the collier his *Davy*, (the name applied in our mines to the safe-lamp,) and he goes to his occupation with the same confidence in this impure atmosphere, that he would do in any other situation, with a candle.

"There has been much quibbling about the *perfect* safety of the wire-gauze lamp. I scarcely know how the words perfect safety can apply to any invention for the preservation of human life; but when we have seen some hundreds of the wire-gauze lamps in daily use for several months past, in all varieties of explosive mixture, in the most dangerous mines of this country, without the slightest accident occurring, it seems only reasonable to infer, that they approximate as nearly to perfect safety as any thing of human contrivance or manufacture can be expected to do.

"It would, however, be quite unreasonable to expect that accidents are never to happen, where the wire-gauze lamps are used; for it must always be remembered, that, setting aside the chance of their being damaged by some of the casualties incidental to coal mining, they are to be entrusted to the management of a body of men amongst whom negligent individuals will be found, who may use damaged lamps, or expose the naked flame to the fire-damp, in spite of the utmost vigilance of the overmen and inspectors of the mines.* Instances of great negligence have occurred, fortunately without any ill consequences—always with the dismissal of the offender from his employment; but it would be absurd to condemn the lamp, or even to quibble upon its want of safety, on this account."

"In the most extensive coal mines† in the North of England where the fire-damp prevails to such an extent, as to require the general use of the safety-lamp, it has been found most advantageous, to employ a steady person to take charge of the lamps, and who is made responsible for keeping them in good order. A chamber is allotted to him, in which he keeps a number of spare lamps, together with oil and cotton for replenishing the lamps which are in use.

"The brass collar of the wire-gauze cylinders are secured to the bottoms of the lamps by locks, which can only be opened by the lamp-keeper, so that the workmen cannot either by accident or carelessness expose themselves to danger by separating the wire-gauze cylinders from the bottoms of the lamps.

"After finishing their day's work, the colliers bring their *Davys* to the lamp-keeper's *cabin*, who unlocking them, takes the bottoms into his own possession, and

* [Mr. George Johnson, "extensively engaged as a viewer in superintending the working of coal mines in the North of England" gave evidence before the select Committee to the same effect. He says, "I think when an accident has happened where Sir H. Davy's lamp has been used, the explosion has been caused, either by some accident happening to and breaking through the wire-gauze, by wilful exposure of the flame to the explosive atmosphere, or by the imperfect state of the wire-gauze." (Minutes, p. 88.) And he relates instances in proof.]

† 2nd Extract, dated May, 1818.

allows the colliers to take the wire-gauze cylinders home for the purpose of cleaning them thoroughly.

"When the colliers return to their work the following morning, the lamp-keeper having replenished the lamps with oil and cotton, lights them and screws on their tops, and then examines them with the utmost care, before he delivers them for use; but if the least injury or defect appears in the gauze, or any other part of a lamp, it is immediately set aside to be repaired, and the person to whom it belongs is supplied with a perfect one.

"After having dispatched the business of the morning, the lamp-keeper occupies himself during the day in walking leisurely through amongst the workmen, carrying some spare lamps with him, to replace such as may happen to be extinguished, &c.

"After a little practice, the lamp-keepers acquire great dexterity in the trimming, &c., of the lamps, and quickly discover the slightest defect or injury in the wire-gauze.

"It is scarcely necessary to observe, that the lamp-keeper's cabin is always placed in a secure part of the mine, as near the workings as circumstances will permit. J. B."

3. *Extract of a Letter from Mr. Peile to Sir H. Davy.*

Colliery Office, Whitehaven,
6th July, 1816.

"I take the liberty of adding a further statement on your invaluable safe lamps, in the Whitehaven collieries belonging to the Earl of Lonsdale, since the first application of them in February last.

"With us, the general use of the lamps, in consequence of the good state of our ventilation, is confined to leading workings, or trial drifts; and in two of these, lately going on in one of the pits unusually infected with fire-damp, and which previously were lighted by means of steel mills, we applied the lamps with great confidence and security.

"In May last in these drifts an extraordinary discharge of fire-damp burst from the pavement of the mine, and the ventilation being at that time unavoidably obstructed, the atmosphere became so charged with fire-damp as to be nearly throughout an explosive mixture. In this situation we derived the unspeakable benefit of light from the lamps, and, notwithstanding the explosive state of the mixture, with the most perfect safety.

"In several other places in the collieries the lamps are used with the same confidence: yet the discharge of fire-damp being moderate, they are not much exposed to explosive mixtures.

"In all the workings shewing the least appearance of fire-damp, the miners are supplied with lamps, and are particularly cautioned to use them on first entering when beginning to work, where, being satisfied of security, they occasionally resort to candles afterwards. This application of the lamp alone, is of the greatest utility, and prevents many slight explosions, and the miners from being burned; besides superseding the necessity of depending on the judgment or discrimination of any individual to prove the existence of the fire-damp, as in the old method, by the candle flame.

"From the repeated proofs made with the lamps, we cannot too strongly express our confidence in their security.

"By experiment, a pint of oil, value six-pence, will about supply a lamp for six days, the ordinary time of a man's working, so that they are cheaper than candles.

"If my humble testimony can in any degree promote the speedy use of the lamp in other places, it will give me great pleasure. J. PEILE."

4. *Extract of a Letter from Mr. John Morris, Plas Issa, 27th Jan.* 1817. *To John Simmons, Esq., Paddington-house.*

"Sir;—You will be pleased to recollect that some time in the month of June last, I applied to you with a request you would send me immediately some of Sir Humphry Davy's safety lamps, in consequence of an explosion of the fire-damp taking place in one of your coal mines, by which several of the men were dreadfully burnt and bruised. On the arrival of the safety lamps, no accurate account of their use accompanied them. But I at length obtained (I think) the Edinburgh Review, in which was a detail of some experiments. This I read to the colliers, which gave them some confidence in the lamps, prior to which they secretly treated them with silent contempt; and I found, notwithstanding these interesting details, that a great doubt existed in their minds. I therefore was obliged to give the most peremptory orders to prepare to descend, and assisting in every preparation and execution myself. But the men's wives, &c. had collected, and made so much noise and lamentations, that it was with some difficulty I could keep them off: having got over this obstacle, and the men down in the pit, instantaneous destruction was momentarily anticipated, when the least noise was heard. I, however, had not the least alarm or the smallest doubt of success, and consequently did all I could to remove their dreadful anxiety. The men had no sooner descended than the enemy was discovered, which they say very much alarmed them, and they would have retreated if they could, but finding that impossible, took courage, and soon found they had destroyed the enemy so far; advancing a little farther, they found him again, and again destroyed him, and so on through the whole work. Thus the first alarm was got over, when all the knowing men in the neighbourhood were got collected together to hear the result, all of which were astonished and amazed, that so simple-looking an instrument should destroy and defy an enemy, heretofore unconquerable. The same precaution and use of the lamp, was gone through the second day, and when the damp was destroyed, we began working and continued to work in this way for some weeks. GEO. MORRIS."

I am possessed of a great number of similar documents respecting the use and application of the lamp;* but they are in general so flattering that I might well be suspected of vanity, if they were to be laid before the public. It has been said that the coal miners have been in many instances disinclined to adopt the safety lamp, and that the proprietors of coal mines have not been always disposed to urge its application. I am anxious that this should be contradicted, for I believe there are no dangerous mines in Great Britain where the safety lamp is not well known, and its efficacy properly estimated, and it is likewise getting into general use on the Continent.

It would be expecting too much from human nature to suppose that there should be no instances of obstinacy in workmen, and of prejudice or indifference in

* [None of these were preserved by him, nor indeed a single letter on the subject; documents of the description above alluded to, seldom, I believe, had his second thoughts.]

coal owners; but these instances have been solitary ones; and if delicacy did not forbid me, I might bring forward numerous proofs of the gratitude and warm feelings with which this invention has been received by the individuals who have benefited by it. I might appeal to the letters of thanks of various individuals, and of the united colliers of Whitehaven, to the vote of thanks of the coal trade of the north of England, of the grand jury of Durham, of the Chamber of Commerce of Mons, and above all, to the present made to me at Newcastle in so flattering a manner, October 11, 1817.*

IV. CONCLUSION.—SOME PRACTICAL OBSERVATIONS.

Time, of course, has increased the confidence of the workmen in the safety lamp; and in a period of nearly two years and a half in the most dangerous mines in Britain, it must have been exposed to all circumstances which the variety of explosive mixtures can occasion. In my first trials with the safety lamp when I found the wire become red-hot, I thought it might be proper to cool it occasionally by water, or moistened cloth; but this was found unnecessary in the common practice of the miners. Whenever a single wire-gauze lamp is made to burn in a very explosive atmosphere at rest, the

* [Vide Vol. I. p. 204. In the same year the medals on Count Rumford's Donation were awarded to the author by the President and Council of the Royal Society, for his labours on the safety lamp and on flame. The figure on these medals is a *free* flame, with the inscription over it, "Noscere quæ vis et causa." In the armorial bearings which were granted when a baronetcy was conferred on him, a flame was introduced, but it was *chained*, with the motto, "Ignis constricta vita secura," which he afterwards changed for that of "Sic semper ardeat."]

heat soon arrives at its maximum, and then diminishes; and the idea of the wires burning out is shown to be unfounded; the carbonaceous matter produced from the decomposition of the oil, tends not only to prevent the oxidation of the metal, but likewise revives any oxide already formed; and this coaly matter, when the fire-damp is burning in the lamp, chokes the upper apertures of the cylinder, and gradually diminishes the heat, by diminishing the quantity of gas consumed.

I have seen wire-gauze lamps in the hands of workmen, which they had used for several months, and which had been often red hot in explosive atmospheres, and which were still perfect.

Where an explosive mixture is in rapid motion, it produces, as has been stated page 77, much more heat: and in this case the radiating or cooling surfaces of the lamp must be increased, or the circulation of air diminished. Twilled gauze, or a double or triple fold of wire gauze on one side of the lamp,* or a screen of metal opposite to the direction of the current, or a semicylinder of glass or of mica within, answers perfectly the object of preventing the heat from rising to redness.

Single iron-wire gauze of the kind used in the common miner's lamp, is impermeable to the flame of all currents of fire-damp, as long as it is not heated above

* [In accordance with the above suggestion, a lamp has been constructed by Mr. Newman which he considers "doubly secure." It is described page 265 of the Minutes of Evidence before the Committee of the House of Commons; it differs chiefly from the common safety lamp in being formed of a double gauze cylinder, the outer a sliding one, and a quarter of an inch apart from the inner. Mr. Newman's evidence respecting the safety of the common lamp is not less strong than that of Mr. Buddle; it is deserving of particular attention, especially in regard to the manner in which Messrs. Upton and Roberts affected to pass flame through it, and which he endeavours to prove (and I think successfully) could not be effected in the foul atmosphere of a mine, only in the open air.]

redness; but if the iron wire be made to burn, as at a strong welding heat, of course it can be no longer safe; and though such a circumstance can perhaps never happen in a colliery, yet it ought to be known and guarded against.

I had an excellent opportunity, September 1816, of making experiments on a most violent blower, at a mine belonging to J. G. Lambton, Esq., [now Earl of Durham.] This blower is walled off from the mine and carried to the surface, where it is discharged with great force. It is made to pass through a leathern pipe, so as to give a stream, of which the force was felt at about two feet from the aperture in a strong current of air. The common single working lamps and double gauze lamps were brought upon this current, both in the free atmosphere and in a confined air. The gas fired in the lamps in various trials, but did not heat them above dull redness, and when they were brought far into the stream they were finally extinguished.

A brass pipe was now fixed upon the blower tube, so as to make the whole stream pass through an aperture of less than half an inch in diameter, which of course formed a most powerful blow-pipe, from which the fire-damp, when inflamed, issued with great violence and a roaring noise, making an intense flame of the length of five feet. The blow-pipe was exposed at right angles to a strong wind, and double gauze lamps and single lamps successively placed in it. The double gauze lamps soon became red hot at the point of action of the two currents; but the wire did not burn, nor did it communicate explosion. The single gauze lamp did not communicate explosion, as long as it was red hot and slowly moved through the currents; but when it was fixed at the point of most intense combustion, it reached a welding heat, the iron wire began to burn with sparks, and the explosion then passed.

In a second and third set of experiments on this violent blow-pipe of fire-damp, single lamps with slips of tin-plate on the outside or in the inside, to prevent the free passage of the current, and double lamps, were exposed to all the circumstances of the blast, both in the open air and in an engine-house where the atmosphere was explosive to a great extent round the pipe, and through which there was a strong current of atmospheric air; but the heat of the wire never approached near the point at which iron wire burns, and the explosion could never be communicated. The flame of the fire-damp flickered and roared in the lamps, but did not escape from its prison.

There is no reason ever to expect a stream of gas of this kind in a mine; but, if it should occur, the mode of reaching and examining it, with most perfect security, is shown; and the lamp offers a resource, which can never exist in a steel-mill, the sparks of which would undoubtedly inflame a current of this kind,

If a workman, having only a common single lamp, finds the temperature of the wire increasing rapidly in an explosive mixture near a blower, he can easily diminish the heat, by turning his back upon the current, and keeping it from playing upon the lamp, by means of his clothes or his body; or by bringing the lamp nearer the orifice, from which the fire-damp issues, he may extinguish it; and there never can be any occasion for him to place his lamp in the exact *point* where two currents, one of fresh air, and one of fire-damp meet each other.

When the fire-damp is inflamed in the wire-gauze cylinders, coal-dust thrown into the lamp burns with strong flashes and scintillations; the miners were at first alarmed by an effect of this kind, produced by the dust naturally raised during the working of the coals. I have made a number of experiments on this subject; but though I have repeatedly thrown coal dust, powdered rosin, and witch meal, through lamps burning in more explosive mixtures than ever occur in coal mines, and though I have kept these substances floating in the explosive atmospheres, and heaped them upon the top of the lamp when it was red hot, yet I never could communicate explosion by means of them. Phosphorus or sulphur are the only substances which can produce explosion, by being applied to the outside of the lamp; and sulphur, to produce the effect, must be applied in large quantities, and blown upon by a current of fresh air.

It will be unnecessary to caution the workmen against heaping sulphur, or gunpowder, or pyrites, which afford sulphur by distillation, upon their lamps; and such dust from these substances as can float in the atmosphere cannot produce inflammation; for minute particles of ignited solid matter have no power of inflaming the fire-damp; and I have repeatedly blown fine coal dust mixed with minute quantities of the finest dust of gunpowder through the lamp burning in explosive mixtures, without any communication of explosion.

A few complaints have been made of the feebleness of the light of lamps after they have been some time used, in consequence of the tissue being choked up by coal dust. But by means of a brush this evil may be removed. And in some experiments, that I made with Mr. Buddle, in the Wall's-end colliery, the light of a single gauze-lamp, furnished with a tin-plate reflector, was found to be superior to that of a common miner's candle, and the light of a lamp without the reflector, nearly equal to it; and almost double to that of the steel-mill at its greatest intensity of light. The trials were conducted by determining the distance at which an object was visible with the different species of light, and considering the intensity of the light directly as the square of the distance.

I have often made lamps, in which surfaces of glass were used for transmitting light without a guard of wire-gauze; they give more light, but they are highly dangerous, and ought never to be used. Thick plates of mica,* (Muscovy glass) may, however, be safely employed, though great care must be taken, in this case, that the radiating and cooling surfaces, where the fire-damp burns, are extensive, and all the precautions mentioned, page 16, must be adopted.

Where a lamp is permanently to be fixed in a part of a mine, it will afford a better light if some of the surfaces are of mica; but for lamps which are constantly to be carried about by the miners, iron wire-gauze, I have no doubt, will be the material always employed. I have tried a lamp on the plan of Argand's, in which wire-gauze feeders were below, and in which a current of air was determined by wire-gauze cylinders above: it gave a brilliant light, but produced so much heat as to boil the oil in the reservoir, and it required a complicated contrivance for trimming it.

* From a very able report on the safety-lamp, drawn up by order of the Chamber of Commerce of Mons, and furnished with some very intelligent and scientific notes, by M Gossart, President of the Chamber, it appears that lamps, with plates of iron, have been used in Flanders. M. Gossart quotes an instance which proves the danger of glass. "A director of the works having descended in the colliery of Tapatouts, with a lamp, of which the base of the cylinder was of glass, a drop of water fell upon and broke the glass, and detached a piece which would have opened a communication for explosion; but the air fortunately, at the moment, was not adulterated with fire-damp."

When a cage of wire of platinum is used within a lamp, even where the explosive mixture burns with flame, it diminishes the heat by its radiation, and it increases the light, so that it will always be useful in lamps; and as it is an imperishable metal, it is only the original expense, which is very small, that is to be attended to. It is proper to urge again what has been mentioned, page 87, that no wire or filament of platinum must be suffered to project beyond the wire gauze, so as to be in the external atmosphere.

The forms of lamps may be infinitely varied; but the most convenient size for a common working lamp is from eight to ten inches in height, and two to two and a half in breadth. If the wire-gauze cylinders are larger, there is too much heat produced in them by the combustion of explosive mixtures.

In gas manufactories, spirit warehouses, or druggists' laboratories where ether is distilled, the common safety lamp may be advantageously used.—A simple mode of exhibiting its power is furnished by throwing a little ether into the bottom of a large jar; the vapour of the ether, mixing with the air, will produce an explosive atmosphere, which will burn within the wire-gauze without inflaming the ether in the bottom of the jar.

If pure hydrogen should be disengaged in any mines, the improbability of which however is very great, wire-gauze of a finer texture must be employed.—This material indeed, it is likely, will be found to possess many new applications, not only in guarding against the communication of flame, but likewise in exerting cooling agencies, wherever elastic media can be exposed to it, so as to have their temperatures lowered by its radiation.

I shall now conclude. Whatever may be the fate of the speculative part of this inquiry, I have no anxiety as to the practical results, or as to the unimpassioned and permanent judgment of the public on the manner in which they have been developed and communicated; and no fear that an invention for the preservation of human life and the diminution of human misery will be neglected or forgotten by posterity.

When the duties of men coincide with their interests, they are usually performed with alacrity; the progress of civilization ensures the existence of all real improvements; and however high the gratification of possessing the good opinion of society, there is a still more exalted pleasure in the consciousness of having laboured to be useful.

APPENDIX.

No. 1.

Since the Researches upon Flame contained in the foregoing pages have been published, M. Gay Lussac has put into my hands a paper written some years ago by M. de Humboldt and himself, which contains some very interesting results that may be adduced as confirmations of my principles, on the causes of combustion and explosion.

MM. Gay Lussac and de Humboldt have shewn that when oxygen and hydrogen are mixed in proportions in which they cannot be fired by the electrical spark, they may be still made to combine in the proportions which can form water, by artificially raising their temperature.

MM. Gay Lussac and de Humboldt suppose that the action of electricity in producing combination is owing to the heat it produces by the compression of the elastic medium through which it passes. This idea is very ingenious, but the phenomena of decomposition by electricity, shew that there is some relation between the primary attractive powers of the chemical elements and their electrical energies.

When the common electrical or voltaic electrical spark is taken in rare air, the light is considerably diminished. I made some experiments to ascertain whether the heat was likewise diminished, and I found that this was certainly the case. Yet in a receiver that contained air sixty times rarer than that of the atmosphere, a piece of wire of platinum, placed in the centre of the luminous arc, produced by the great voltaic apparatus of the Royal Institution, became white hot; and that this was not owing to the electrical conducting powers of the platinum, was proved by repeating the experiment with a filament of glass, which instantly fused in the same position.

It is evident from this, that electrical light and heat may appear in atmospheres in which the flame of combustible bodies could not exist, and the fact is interesting from its possible application in explaining the phenomena of the Aurora Borealis and Australis.

No. 2.

M. Sementini, professor of chemistry at Naples, presented me in 1819 with a lamp, in which alcohol burnt without flame, by means of fine coils of silver wire, and afforded phenomena exactly of the same kind, as the lamp furnished with wire of platinum.

When I first discovered the phenomena of the ignition of thin filaments, of platinum and palladium, I ascertained that the temperature required for this result was much below ignition; but I did not determine the precise degree on Fahrenheit's scale. It was evident, however, from the principles laid down at page 76, that it must be lower for hydrogen than most other inflammable gases; and lower in proportion as the wire or foil was finer.

M. Dobreiner has lately made the discovery that the finely divided and spongy platinum, obtained by precipitation from solution, becomes ignited, even at common temperatures; and MM. Thenard, and Dulong, and other chemists, in repeating his observation, have found that various metals in a finely divided state, have the same property of hastening or producing combinations at a lower temperature, than those at which they usually occur, and have given many facts analogous to those described page 84. Mr. Garden has found the ore of iridium likewise possessed of the property of inflaming mixtures of common air and hydrogen.

It is probable that the rationale of all these processes is of the same kind. Whenever any chemical operation is produced by an increase of temperature, whatever occasions an accumulation of heat, must tend to give greater facility to the process; a very thick wire of platinum does not act upon a mixture of oxygen and hydrogen, at a heat below redness; but if beat into thin laminæ, it occasions its combustion at the heat of boiling mercury, and, when in the form of the thinnest foil, at usual temperatures. I cooled the spongy platinum to 3° of Fahr., and still it inflamed hydrogen nearly of the same temperature, issuing from a tube cooled by salt and ice.

I thought that common radiant heat or light, might be necessary to the effect; but the cooled metal and the gases acted with the same phenomena in darkness.

It may be supposed that the spongy platinum absorbs

hydrogen, or that it contains oxygen; but neither of these hypotheses will apply to the fact that I first observed, of the ignition of fine wires in different mixtures of inflammable gases and air, at temperatures so far below ignition.

A *probable* explanation of the phenomenon, may, I think, be founded upon the electro-chemical hypothesis which I laid before the Royal Society in 1806; and which has been since adopted and explained, according to their own ideas, by different philosophers.

Supposing oxygen and hydrogen to be in the relations of negative and positive, it is necessary to effect their combination, that their electricities should be brought into equilibrium or discharged. This is done by the electrical spark or flame, which offers a conducting medium for this purpose, or by raising them to a temperature, in which they become themselves conductors. Now platinum, palladium, and iridium are bodies very slightly positive with respect to oxygen; and though good conductors of electricity, they are bad conductors and radiators of heat, and supposing them in exceedingly small masses, they offer to the gases the conducting medium necessary for carrying off, and bringing into equilibrium their electricity without any interfering energy, and accumulate the heat produced by this equilibrium. Other metals do not possess the same union of qualities, yet most of them assist combination at lower temperatures than glass, which is a non-conductor of electricity.

That spongy platinum, even when moistened, as M. Dobreiner has very lately shown, should facilitate the combination of oxygen and hydrogen, *may* depend upon *this peculiar* electrical property; and why foil of platinum should have its power of causing oxygen and hydrogen to combine, increased by being placed, for a short time, in nitric acid, as MM. Dulong and Thenard have shown, may be owing to this, that the slight positive charge it acquires may, in being brought into equilibrium, be a first step in the operation: and there are analogous instances.*

Fine wire of platinum, I find, when conveying currents of electricity, as in a circuit, with zinc and sulphuric acid, or charcoal and nitromuriatic acid, has not its power of acting upon gaseous mixtures sensibly increased.

No. 3.

General inflammable air is only disengaged in coal-mines; yet the salt works of Styria, Salzburg, and Upper Austria are not exempt from accidents depending upon carbonated hydrogen gas. An explosion had happened at Aussee, in 1818, a few weeks before I visited the salt works, by which several persons were killed; and the miners received the safety-lamp, and witnessed its operation with gratitude and surprise.

The inflammable air appeared to me, in these instances, to be derived from bituminous schist.

No. 4.

I have had some correspondence with Mr. Buddle respecting the accidents which have happened in coal mines, since the discovery of the safety-lamp. He refers them, in all cases, to the carelessness of workmen.

I should strongly recommend double lamps, in cases

* [Vide the later results of the experiments of Mr. Faraday on this curious subject, (Phil. Trans. for 1834) many of which accord with the above explanation.]

where miners are obliged to work for any time in explosive mixtures, or wherever currents are expected;—or lamps with mica, or tin-plate *within* the wire-gauze to prevent too great a circulation of air, (see p. 103.) It is very easy to extinguish a lamp in which the fire-damp is burning, by sliding a tin-plate cylinder over it, or by a circle of wire-gauze fitting the interior in a rim of copper and moved by the termination of the trimming wire: but it is much better, in all cases of danger, to use lamps which, *under no circumstances*, can explode. Such as those described in p. 77.

Having often trusted my life to the safety-lamp under the most dangerous circumstances, I cannot but sometimes smile when the public papers endeavour to invalidate its security upon the opinions or evidence of certain persons who have their own nostrums for preventing the accumulation of inflammable air in mines.

I have sometimes to read letters on the improvement of the invention, by plans, most of which are discussed in the foregoing pages; such as using glass or mica as a part of the surface for transmitting light, using double lamps, or double lamps containing a reflecting surface to prevent explosions from currents; and I have actually seen a lamp upon the rudest model of those I first made, having thick glass above, and wire-gauze below, called "The newly invented Safety Lamp."*

No. 5.

For gas manufactories, or houses where gas is extensively used, I should recommend the safety-lamp with iron wire-gauze, but for the use of the navy, those with copper wire-gauze are less liable to rust. As the

* [This, I apprehend, was Messrs. Upton and Roberts' lamp as it first appeared.]

latest instance of a ship lost, for want of a safety-lamp, I may mention the "Kent" East Indiaman, which was burnt, as I am informed by the Shipping Committee, in consequence of the inflammation of rum, by means of a common lantern.

[As, since the death of the Author, from time to time, the efficacy of the safety-lamp has been called in question, in a very injurious manner in relation to the public good; and further, as claims have occasionally been set up, which if established would detract from the merit of its inventor; I think it right to adduce some evidence respecting both; first, in proof, that the instrument used with the precautions pointed out is truly a safety-lamp and deserving of all confidence; and next, that the Author alone is justly entitled to be considered its discoverer. With this double object in view, I shall first have recourse to the statements of Mr. Buddle made before a Select Committee of the House of Commons, on accidents in mines, already referred to. The passages I shall select (and it is necessary to make a selection, as his ample evidence covers forty-seven folio pages,*) will relate principally to the discovery of the lamp,—the causes of danger in its use, and the means of guarding against them; circumstances which cannot be too much dwelt upon, and which in Mr. Buddle's plain and animated narrative are so represented as to make a very forcible impression.

* Vide "Report from the Select Committee on Accidents in Mines; together with minutes of evidence and index, ordered by the House of Commons to be printed, 4th September, 1835." The Report occupies 8 folio pages; the minutes of evidence 320 pages, and the index 41 pages.

To the question of the Committee, "Have you ever formed any experiment upon the Davy lamp, with a view of ascertaining what force is necessary to compel the passage of the flame through the gauze?" Mr. Buddle replied, "I have not by myself, but in company with Sir Humphry Davy; and I am very happy to have this opportunity of doing justice to the memory of my lamented friend Sir Humphry Davy on this particular point. The lamentable accidents which, prior to the year 1815, had occurred in our neighbourhood naturally directed the attention of all humane persons to the subject. Sir Humphry Davy was introduced to me by the late Bishop of Bristol, and he called upon me at the Walls-End colliery one day to inquire into the nature and cause of this lamentable catastrophe; I explained to him as well as I was able the nature of our fiery mines, and that the great desideratum was a light that could be safely used in an explosive mixture. I had not the slightest idea myself of ever seeing such a thing accomplished. After a great deal of conversation with Sir Humphry Davy, and he making himself perfectly acquainted with the nature of our mines, and what was wanted, just as we were parting he looked at me and said, 'I think I can do something for you.' Thinking it was too much ever to be achieved, I gave him a look of incredulity; at the moment it was beyond my comprehension. However, smiling, he said, 'Do not despair; I think I can do something for you in a very short time.' I should think, to the best of my recollection, within fourteen days he wrote to me to say that he flattered himself, he had done the thing,—that he had made a discovery which would answer my object; namely, the procuring a safe light in an explosive mixture. In a few days he sent me down two of the Davy lamps, as nearly as possible like that before the Committee; but I have one of our working lamps with me. He told me that it would burn safely in an explosive mixture; that there was no hazard except in exposing it to a strong current, by which the explosion would be passed through the gauze cylinder; he, therefore, cautioned me particularly against such an exposure; but he had a remedy for it. On the strength of his authority I took this lamp without hesitation into an explosive mixture. I first tried it in an explosive mixture on the surface, and then took it into a mine; and, to my astonishment and delight, it is impossible for me to express my feelings at the time when I first suspended the lamp in the mine, and saw it red hot; if it had been a monster destroyed, I could not have felt more exultation than I did. I said to those around me, 'We have at last subdued this monster.' Sir Humphry, some time afterwards came down to the North; he asked me if we had situations under-ground where he could see the effect of his lamp. I told him I could show him in the pit every degree of inflammability in all stages, from the bare firing point up to the most inveterate mixture, the most highly explosive,—that in fact would extinguish the lamp by explosion within the gauze cylinder. He went down this very pit with me in which the accident happened lately, the G. pit at Walls-End colliery. I took him through all the various stages of explosive mixture; I showed him all the gradations from the *thickening* of the air which produces a little elongation of the flame from a small admixture of inflammable air, to where we were working pillars, and where we were exposed to large quantities of gas from the goaves. There we took the lamp and suspended it for a length of time, till it was red-hot, during which, at the same time, we exposed it to the ordinary impulse of the current of air, (there were currents of different degrees of inflammability); and in different parts we tried the lamp in a red-hot state, for a considerable length of time. I was present, and the under-viewer, the overmen, and others were also present."

2227. "Can you define the time in hours and minutes?"—"No, I cannot, but I should suppose from a quarter of an hour to half an hour, during which Sir Humphry was explaining to us the nature of the lamp; it was a lecture in fact upon the spot, and how the time might pass during that interval I cannot tell, but I think we might be down the mine for a period of two or three hours altogether. He then explained to us the danger of exposing the lamp to a strong current of gas, or even to a strong current of explosive mixture, as it would risk the passing of the flame through the gauze, but he pointed out a remedy at the same time for that contingency, and which we have always used; namely, by a tin screen, which slides upon the frame-wires of the lamp, and encircles the circumference of the gauze cylinder to an extent of about one-half to two-thirds of its circumference. He could not there show us the effect of the passing of the explosion to make us sensible of the danger to which we were exposed in that way, and teach us how to avoid it. But some time previously to that, I accompanied him to one of the Earl of Durham's collieries,—to what is called the Morton West pit, where a very large blower from the shaft, not from the coal but from a fissure in the stone, had been for many years discharging up the shaft, by a cast-iron pipe to the surface, in a similar manner to that which I have already described at

Walls-End. We took a length of hose from an extinguishing engine, with the jet-pipe upon it, and attached that to the blower-pipe at the top of the pit; it was held horizontally, and the jet was thrown very forcibly out of the nozzle of the pipe; the blower was sufficiently strong to propel the stream of gas across the engine-house. I well recollect the pipe was held at the entrance of the engine-house, and the jet passed the explosion nearly to the far end of the room, for it was very powerful; the distance that the blower fired it was from nine to twelve feet I should think. I held the lamp in the direction of the jet, and not having seen it before, I was not very apprehensive of its firing. It did not fire at first, but as I approached the end of the nozzle-pipe, the gauze became heated red-hot and passed the explosion. The flame was as long or longer than the breadth of the engine-room; I remember that it burnt the nap off my great coat and spoiled it. This experiment was repeated over and over again. Lord Durham himself was present, and a great many other persons, professional men and others, were present on this occasion. The force of Sir Humphry's remarks at the time was 'Now, gentlemen, you see the nature of the danger to which you are exposed in using the lamp, and I caution you to guard against it in the manner I have shown you. This is to show the only case in which the lamp will explode, and I caution and warn you not to use it in any such case when you can avoid it without using the shield.' In fact, this is a kind of test to which the lamp never need be exposed in practice; we never have occasion to approach the mouth of a blower so close as to expose ourselves to the blast of the issuing gas. Blowers are very different from pipes, as instead of issuing out of a single orifice, as from the jet of the extinguishing engine, they generally issue from a long crack or *thread* of indefinite extent; sometimes the thread may be a foot or a yard long, and occasionally I have seen them five yards and more long; when fired they generally form a sheet of fire, rather than a jet like a gas-pipe. In fact, I do not think I ever saw the orifice of a blower of any magnitude that was so small that I could not put my finger or even my fist into it; some parts of the crack or blower threads are close for a foot or two, and then open and close again; and if you set all the orifices of one blower-thread on fire, you may have it burning out in different places at the same time, but none approaching to what you would call the end or nozzle of a pipe. The orifices are not circular, and consequently the jets from them are diffused and do not form a concentrated blast. It cannot be necessary in any case to expose the Davy Lamp to the direct jet of gas from a blower. The only case in which I conceive danger can arise by exposing the Davy to the impulse of an explosive current is in travelling the gas-pipe drifts. Explosive mixtures occasionally occur there, when the current is moving with a velocity of from three to four feet per second, which is a palpable breeze, and sensibly biases the flame within the wire-gauze cylinder; in fact, it propels the flame against one side of the cylinder."

"Is the Committee to understand that that is when the lamp is carried, or when it is stationary in foul draught?"—"Either in the one case or the other."

"You make no difference when the lamp is carried against the stream?"—"Yes, that causes a double impulse; if the person carrying the Davy is going with the same velocity to windward as the air is moving; but we use the lamp in this manner as little as we can; our rule is to travel with the current of air as much as possible. All the people employed in this occupation, the waste-men, have shields upon their lamps, to shelter and prevent their being affected by the current; and, moreover, they carry their lamps in a particular manner when they are travelling either with or against the current, by which they shelter them by their bodies, and the shield is always opposed to the direction of the current."

"Why is not the shield made of a transparent substance?"—"We do not find it necessary; the tin shield in some degree answers the purpose of a reflector, and the cheapness of the tin is also an object."

"If it was a transparent substance it would give light upon that side as well when the blast came?"—"Yes; but anything in the shape of a lantern or lamp is much better not to give light on both sides, and it shows a more agreeable light with an opaque shield, especially when you are travelling in numbers together. The naked Davys do not afford so agreeable a light in travelling as those with shields in my opinion."

"The tin shields do not dazzle the eye so much?" "They do not."

"Though the danger is imminent in travelling those wastes, the men are so accustomed to it, and understand the theory as well as the practice of using the Davys so well, that you never knew an accident occur to them?" —"I never did. I can only say, that from the time of the invention of this lamp, and getting them introduced as quickly and as extensively as possible, I am quite certain that I have not for many years had less than 1000 lamps a day using, and frequently 1500, and I can state that I never have known an explosion happen from the Davy lamp; not even in one solitary instance. I have seen them in operation, and been with them myself in all possible varieties of explosive mixtures to which they can be exposed in our mines. From my experience I have perfect confidence in them; and when I state, during the number of years (nearly twenty) which I have had those lamps in daily use, that I do not know of one single accident having happened from them, at least as far as my own experience goes, I think I am warranted in pronouncing the Davy lamp to approximate as nearly to perfection as any instrument of human invention can do. This is my decided opinion."

Many other parts of Mr. Buddle's evidence before the Committee, might be brought forward with advantage, were I not restricted as to space. The curious reader, and the reader really interested in the subject I must refer to the Minutes themselves for full satisfaction; I must refer him too, to the same document for the opinions of the other individuals examined,* which altogether are exceedingly miscellaneous, and of different degrees of value. I shall avoid comment on them on many accounts, as well as on the various imitative safety lamps industriously brought, or attempted to be forced, into notice.

From the Report itself, I shall give an extract, partly on account of the acknowledgment expressed in it, of the benefits derived from the safety-lamp, and partly on account of a misapprehension relative to the principle of the invention. The passage is as follows:—

"Your committee have endeavoured to investigate,

* [It is satisfactory to find that the highest practical authority relative to the safety-lamp, viz. Mr. Buddle's, is supported by the testimony of the majority of witnesses practically acquainted with the use of the lamp, and who give testimony on their own experience, as Anthony Winship, Ralph Elliott, Messrs. G. Johnson, G. Mitcheson, R. Smith, J. Garforth, T. Embleton.]

with strict impartiality, the merits of the different lamps which have been brought under their notice. In the course of the evidence many varieties will be found described. The invention claimed by the late Sir Humphry Davy, on principles demonstrated by that able philosopher, may be considered as having essentially served the mining interests of this kingdom, and through them contributed largely to the sources of national as well as individual wealth. Many invaluable seams of coal never could have been worked without the aid of such an instrument; and its long use through an extensive district, with the comparatively limited number of accidents, favours its claim to be considered, under ordinary circumstances, a safety-lamp. The principles of its construction appear to have been practically known to the witnesses, Clanny and Stephenson, previously to the period when Davy brought his powerful mind to bear upon the subject, and produced an instrument which will hand down his name to the latest ages."

The latter part of this passage is the misapprehension I have alluded to. Were it not so,—were it correct that Dr. Clanny and Mr. Stephenson had anticipated the author in the discovery of the principle of the safety-lamp, theirs would have been the merit, not his, of originality; and theirs should have been the honour and the reward. In proof of the misapprehension, I shall insert some documents, already published, at an earlier period, when similar claims to priority of discovery were as unjustly as injudiciously made by the friends of these gentlemen. The first document I shall give is a letter from the author, relative to the principle of his lamp and that of Dr. Clanny. It was addressed to the Reverend Dr. Gray.

"23, *Grosvenor Street*,
Dec. 13, 1815.

"My Dear Sir,

"A friend of mine has sent me a newspaper, the Tyne Mercury, containing a very foolish libel upon me. It states, amongst other things, that I did not mention Dr. Clanny or his lamp in my late paper read before the Royal Society; whereas I mentioned his lamp as a very ingenious contrivance, and named him amongst the gentlemen who obligingly furnished me with information upon the subject.

"It will be needless for me to point out to you that my lamp has not one principle in common with that of Dr. Clanny. He forces in his air through water by bellows. In mine the air passes through safety canals without any mechanical assistance. Mine is a common lantern made close, and furnished with safety canals.

"I hope I shall not hear that Dr. Clanny has in any way authorized or promoted so improper a statement as that in the Tyne Mercury; indeed I do not think it possible.

* * * * *

"I am, my dear sir,
"Your sincerely obliged
"H. Davy."

The two documents I shall give next relate to Mr. Stephenson; they are the resolutions of men of the highest respectability, in each instance assembled expressly for the purpose of investigating the claims set up in favour of his invention.

At a general meeting of coal owners, held on the 26th November 1817, at which the Earl of Durham presided, it was agreed,

"That the Resolution passed at the meeting of the friends of Mr. G. Stephenson, on the 5th inst., impugn the justice and propriety of the proceedings of a meeting of the coal-trade on the 31st of August 1816.

"That the present meeting feel themselves called upon, as an act of justice to the character of their great and distinguished benefactor Sir Humphry Davy, and as a proof that the coal-trade of the north in no way sanctions the resolutions of Mr. Stephenson's friends, to state their decided conviction, that the merit of having discovered the fact, that explosions of fire-damp will not pass through tubes and apertures of small dimensions, and of having applied that principle to the construction of a safety-lamp, *belongs to Sir Humphry Davy alone.*

"That this meeting is also decidedly of opinion, from the evidence produced in various publications by Mr. George Stephenson and his friends, subsequently to the meeting of the coal trade, which was held on the 18th March 1816, as well as from the documents which have been read at this meeting, that Mr. Stephenson *did not* discover the fact, that explosions of fire-damp will not pass through tubes and apertures of small dimensions; and that he *did not* apply that principle to the construction of a safety-lamp; and that the latest lamps made by Mr. Stephenson are evident imitations of those of Sir Humphry Davy, and that even with that advantage, they are so imperfectly constructed as to be actually unsafe.

"That the above resolutions be published thrice in the Newcastle papers, and in the Courier, Morning Chronicle, and Edinburgh Courant; and that printed copies thereof be sent to the Lords Lieutenants of the two counties, to the Lord Bishop of Durham, and to the principal owners and lessors of collieries upon the Tyne and Wear."

The following resolutions are those of men of science, come to at a meeting held for "considering the facts relative to the discovery of the safety-lamp."

"*Soho Square, November 20th,* 1817.

"An advertisement having been inserted in the Newcastle Courant of Saturday, November 7th, 1817, purporting to contain the resolutions of 'A meeting held for the purpose of remunerating Mr. George Stephenson for the valuable services he has rendered mankind by the invention of his safety-lamp, which is calculated for the preservation of human life in situations of the greatest danger.'

"We have considered the evidence produced in various publications, by Mr. Stephenson and his friends, in support of his claims; and having examined his lamps and inquired into their effects in explosive mixtures, are clearly of opinion—

"First, That Mr. George Stephenson *is not* the author of the discovery of the fact, that an explosion of inflammable gas will not pass through tubes and apertures of small dimensions.

"Secondly, That Mr. George Stephenson *was not* the first to apply that principle to the construction of a safety-lamp, none of the lamps which he made in the year 1815 having been safe, and there being no evidence even of their having been made upon that principle.

"Thirdly, That Sir Humphry Davy not only discovered, independently of all others, and without any knowledge of the unpublished experiments of the late Mr. Tennant on flame, the principle of the non-communication of explosions through small apertures, but that he also has the sole merit of having first applied it to the very important purpose of a safety-lamp, which has evidently been imitated in the latest lamps of Mr. George Stephenson.

(Signed) " JOSEPH BANKES, F.R.S.
" WM. THOMAS BRANDE.
" CHARLES HATCHETT.
" WM. HYDE WOLLASTON."

The testimony in these documents, was, at the time, I believe, considered conclusive, and as nothing has transpired since to weaken the force of it, it must be considered equally so at present by all unbiassed judges. Why, in forming and giving their opinion, the select committee did not refer to it in their Report, especially as Dr. Clanny and Mr. Stephenson, the parties concerned, were summoned as witnesses, it is difficult to imagine.

Perhaps, it may be said, that I have not justly interpreted the latter part of the paragraph, which I have quoted from the Report; and that I have attached too much importance to the expression declarative of these gentlemen having been practically acquainted with the principle of the safety-lamp before it was discovered by the author. This I cannot admit. If practical knowledge of a principle has any just meaning, it includes theoretical knowledge of it. An artist may be acquainted with a process, and ignorant of its principle, but he cannot be practically acquainted with its principle without knowing its principle; it involves a contradiction. The misapprehension referred to in the Report is of this kind. I can readily believe that the members of the committee, or the individual by whom the Report was drawn up, did not wish to detract from the merit of the inventor of the safety-lamp, and had no intention of bringing forward as rivals, in regard to originality of invention, either Dr. Clanny or Mr. Stephenson; but, assuredly, the logical deduction from their expression is detractive, and even more so, in assigning to these gentlemen, as it does, a priority of knowledge of the principle, although their lamps and the safety-lamp of the author had nothing in common involving principle. Their object indeed was the same, viz., to confine flame, as must necessarily be the aim of every kind of safety-lamp attempted, in which the source of light is flame; but what has this to do with identity of principle? In further illustration of the mistaken view in the Report, I shall give a few extracts from the Minutes of Evidence, expressing Mr. Stephenson's opinion regarding the principles of the lamps, and Dr. Clanny's.

I shall limit myself to one extract from Dr. Clanny's evidence; it is in reply to a leading question put to him by the Committee relative to the principle of the wire-gauze safety-lamp, compared with his own original lamp, in which security was ingeniously attempted by transmitting the air feeding the flame through water, and so isolating it.

"340. Sir Humphry Davy's wire-gauze lamp was constructed upon a like principle of isolation, but was admitted to be a simplification of the application of that principle?"

"By many it certainly was, and by me (Dr. Clanny) amongst the rest, as regards isolating a light."

Let us now turn to Mr. Stephenson's opinion.

"1547. Do you (Mr. Stephenson) consider Sir Humphry Davy's views of the efficiency of the lamp constructed by him to have been based upon mechanical or chemical principles."

"On chemical principles."

"1548. And your own on mechanical principles?"

"Certainly, entirely."

"1561. Do you (Mr. Stephenson) not think that Dr. Clanny exhibited the principle which has been employed by yourself and Sir Humphry Davy?"

"The principle is not the same; but I think he was the first person who invented a lamp to be taken into an explosive mixture."

Before concluding, I find it necessary to advert to another part of the Parliamentary Report. The Committee referring to certain experiments which were made before them in the Laboratory of the London University, by Mr. Pereira, on different forms of the wire-gauze safety-lamp, offer the following remarks.

"In the experiments made before your Committee at the London University, it may possibly be remarked, that the tests applied were not such, in nature or mode of application, as the known actual condition of the mines would point out as satisfactory. It must not be forgotten that the object of those experiments was to ascertain which, of all the lamps produced, was, when exposed to the severest trial, best entitled to the name of a "Safety-lamp." In these experiments the explosion of the gases within the lamp was effected in every one, and similar explosions produced externally, save Messrs. Upton and Roberts'. Your Committee are, therefore, decidedly convinced that its construction possesses paramount merit. Your Committee cannot admit that these experiments had any tendency to detract from the character of Sir H. Davy, or to disparage the fair value placed by himself upon his invention. The improvements are probably those which longer life and additional facts, would have induced him to contemplate as desirable, and of which, had he not been the inventor, he might have become the patron."*

Relative to the experiments thus alluded to, I may remark, that the objection deduced from them against the safety-lamp, cannot be considered valid, inasmuch as the Author never intended that it should be subjected to a test, not applicable to it in the mines for which it was designed. The explosions alluded to, were produced by artifice and management, and under circumstances which cannot possibly concur in an explosive atmosphere in a colliery.

Relative to the improvements required, imagined by the Committee and Messrs. Upton and Roberts' patent lamp, so highly lauded, and which hitherto appears only to have been tested in the Laboratory, I beg to offer the opinions of Mr. Buddle.

"From the period when Sir Humphry Davy first produced his lamp to the time of his death, had you any communications with him?"

"Very frequently; not only communications by letters, but I also visited Sir Humphry Davy when I came to town, and he used to call upon me when he came to the North."

"Did he ever express the hope that he should make improvements upon that lamp?"

"No; in fact I was so perfectly satisfied with what he had done, that I did not ask for any thing more; from its simplicity and portability it amounted to every thing I could desire. The travellers, for instance, the wastemen, whose business it is to work and travel in the wastes for six hours daily, with small intervals of rest, sitting down occasionally at their halting-places, — they are frequently travelling through places not above three

* Report from the Select Committee on Accidents in Mines, viii.

feet or four feet high (and sometimes creeping) in a very uneasy attitude. None can do this sort of work but men in the constant habit of it, and they are obliged to sit down now and then to rest. Carrying this lamp for six hours in so constrained an attitude, renders it important that the lamp should be as light and portable as possible; and I question the ability of the strongest man to carry a lamp of that size for six hours, (*pointing to a lamp of greater size on the table of the Committee*)." Minutes, p. 159.

Since this inquiry was instituted, little more than four years have elapsed, and the lamp just alluded to is almost forgotten; I have been informed by Mr. Buddle, that it has been found totally inapplicable for use in the "fiery collieries," its light being extinguished on entering an explosive atmosphere,—the situation in which a safety-lamp is mainly required.]

II.

SOME EXPERIMENTS AND OBSERVATIONS ON THE COLOURS USED IN PAINTING BY THE ANCIENTS.*

Introduction.

THE importance the Greeks attached to pictures, the estimation in which their great painters were held, the high prices paid for their most celebrated productions, and the emulation existing between different states with regard to the possession of them, prove that painting was one of the arts most cultivated in ancient Greece: the mutilated remains of the Greek statues, notwithstanding the efforts of modern artists during three centuries of civilization, are still contemplated as the models of perfection in sculpture, and we have no reason for supposing an inferior degree of excellence in the sister art amongst a people to whom genius and taste were a kind of birth-right, and who possessed a perception, which seemed almost instinctive, of the dignified, the beautiful, and the sublime.

The works of the great masters of Greece are unfortunately entirely lost. They disappeared from their native country during the wars waged by the Romans, with the successors of Alexander, and the later Greek Republics, and were destroyed either by accident, by

* [From the Phil. Trans. for 1815.—Read before the Royal Society, Feb. 23, 1815.]

time, or by barbarian conquerors, at the period of the decline and fall of the Roman Empire.

The subjects of many of these pictures are described in ancient authors: and some idea of the manner and style of the Greek artists may be gained from the designs on the vases, improperly called Etruscan, which were executed by artists of Magna Græcia, and many of which are probably copies from celebrated works: and some faint notion of their execution and colouring may be gained from the paintings in fresco found at Rome, Herculaneum, and Pompeii.

These paintings, it is true, are not properly Greek; yet whatever may be said of the early existence of painting in Italy as a national art, we are certain, that at the period when Rome was the metropolis of the world, the fine arts were cultivated in that city exclusively by Greek artists, or by artists of the Greek schools. By comparing the descriptions of Vitruvius* and Pliny with those of Theophrastus,† we learn that the same materials for colouring were employed at Rome and at Athens; and of thirty great painters that Pliny mentions, whose works were known to the Romans, two only are expressly mentioned as born in Italy, and the rest were Greeks. Ornamental fresco painting was indeed generally exercised by inferior artists; and the designs on the walls of the houses of Herculaneum and Pompeii, towns of the third or fourth order, can hardly be supposed to offer fair specimens of excellence, even in this department of the art: but in Rome, in the time of her full glory, and in the ornaments of the imperial palace of the first Cæsars, all the resources of the distinguished artists of that age were probably employed.

* De Architectura, lib. vii. cap. v.

† De Lapidibus.

Pliny names Cornelius Pinus, and Accius Priscus, as the two artists of the greatest merit in his own time, and states, that they painted the Temple of Honour and Virtue,* — "Imperatori Vespasiano Augusto restituenti:" and it is not improbable that these artists had a share in executing or directing the execution of the paintings and ornaments on the baths of Titus; and at this period the works of Zeuxis, Parrhasius, Timanthes, Apelles, and Protagoras, were exhibited in Rome, and must have guided the taste of the artists. The decorations of the baths were intended to be seen by torchlight, and many of them at a considerable elevation, so that the colours were brilliant and the contrast strong; yet still the works are regarded by connoisseurs as performances of considerable excellence: the minor ornaments of these have led to the foundation of a style in painting which might, with much more propriety, be called Romanesque than Arabesque: and no greater eulogy can be bestowed upon them than the use to which they have been applied by the greatest painter of modern times, in his exquisite performances in the Vatican. In these, and in other works of the same age, the effects of the ancient models is obvious; and the various copies and imitations that have been made of these remains of antiquity, have transferred their spirit into modern art, and have left little to be desired as to those results which the skill of the painter can command. There remains, however, another use to which they may be applied, that of making us acquainted with the *nature* and *chemical composition* of the colours used by the Greek and Roman artists. The works of Theophrastus, Dioscorides, Vitruvius, and Pliny, contain descriptions of the substances used by the ancients as pig-

* Plin. Nat. Hist. lib. xxxv. cap. 10.

ments: but hitherto, I believe, no experimental attempt has been made to identify them, or to imitate such of them as are peculiar.* In the following pages I shall have the honour of offering to the Society an investigation of this subject. My experiments have been made upon colours found in the baths of Titus, and the ruins called the baths of Livia, and in the remains of other palaces and baths of ancient Rome, and in the ruins of Pompeii. By the kindness of my friend, the celebrated Canova, who is charged with the care of the works connected with ancient art in Rome, I have been enabled to select, with my own hands, specimens of the different pigments that have been found in vases discovered in the excavations lately made beneath the ruins of the palace of Titus, and to compare them with the colours fixed on the walls, or detached in fragments of stucco: and Signor Nelli, the proprietor of the Nozze Aldovrandine, with great liberality permitted me to make such experiments upon the colours of this celebrated picture, as were necessary to determine their nature. When the preservation of a work of art was concerned, I made my researches upon mere atoms of the colour, taken from a place where the loss was imperceptible: and without having injured any of the precious remains of antiquity, I flatter myself I shall be able to give some information,

* In the 70th volume of the Annales de Chimie, page 22, M. Chaptal has published a paper on seven colours, found in a colour shop at Pompeii. Four of these he found to be natural colours, ochres, a specimen of Verona green, and one of pumice stone. Two of them were blues, which he considers as compounds of alumine and lime, with oxide of copper, and the last a pale rose colour, which he regards as analagous to the latter, formed by fixing the colouring matter of madder upon alumine. I shall again refer to the observations of M. Chaptal in the course of this paper. It will be found, on perusal, that they do not supersede the inquiry mentioned in the text.

not without interest to scientific men, as well as to artists, and not wholly devoid of practical applications.*

2. *Of the Red Colour of the Ancients.*

Amongst the substances found in a large earthen vase filled with mixtures of different colours with clay and chalk, found about two years ago in a chamber at that time opened in the baths of Titus, are three different kinds of red:—one bright, and approaching to orange; another dull red; a third a purplish red.† On exposing the bright red to the flame of alcohol, it became darker red; and, on increasing the heat by a blow-pipe, it fused into a mass, having the appearance of litharge, and which was proved to be this substance, by the action of sulphuric and muriatic acids. This colour is consequently minium, or the red oxide of lead.

On exposing the dull red to heat, it became black; but on cooling, recovered its former tint. When heated in a glass tube, it afforded no volatile matter condensible by cold but water. Acted on by muriatic acid, it rendered it yellow; and the acid, after being heated upon it, yielded an orange-coloured precipitate to ammonia. When fused with hydrate of potassa, the colour rendered it yellow; and the mixture acted on by nitric acid afforded silica and orange oxide of iron. It is

* [In the Memoirs of the Life of the Author, pages 185 and 444, his taste for the fine arts is mentioned; of which the introduction to this paper might be adduced, were it necessary, as an additional proof; I say were it necessary, because it can hardly be supposed that a person of quick sensibility as he was, of a lively and poetical imagination, an ardent admirer of all that is beautiful and impressive in Nature, could fail to be an admirer of the great works of art: the same feelings are touched by both,—both mainly seem to act on the mind in a similar manner; and is not this because the elements of both are similar ?]

† Nearly of the same tint as prussiate of copper.

evident from these results, that the dull red colour is an iron ochre.

The purplish red submitted to experiments, exhibited similar phenomena, and proved to be an ochre of a different tint.

In examining the fresco paintings in the baths of Titus, I found that these colours had been all of them used—the ochres particularly—in the shades of the figures, and the minium in the ornaments on the borders.

I found another red on the walls, of a tint different from those in the vase, and much brighter, and which had been employed in various apartments, and formed the basis of the colouring of the niche, and other parts of the chamber, in which the Laocoon is said to have been found. On sweeping a little of this colour from the wall, and submitting it to chemical tests, it proved to be vermilion or cinnabar; and, on heating it with iron filings, running quicksilver was procured from it. I found the same colour on some fragments of ancient stucco in a vineyard, near the pyramidal monument of Caius Cestius.

In the Nozze Aldovrandine, the reds are all ochres. I tried on these reds the action of acids, of alkalies, and of chlorine, but could discover no traces, either of minium or vermilion, in this picture.

Minium was known to the Greeks under the name of *σανδαράχη*,* and to the Romans, under that of cerussa usta. It is said by Pliny† to have been discovered accidentally by means of a fire that took place at the Piræeus at Athens. Some ceruse, which had been exposed to the fire, was found converted into minium;

* Dioscorides, lib. v. 122.

† Pliny, lib. xxxv. cap. 20.

and the process was artificially imitated: and he states that it was first used as a pigment, by Nicias.*

Several red earths used in painting are described by Theophrastus, Vitruvius,† and Pliny. The Sinopian earth, the Armenian earth, and the African ochre, which had its red colour produced by calcination.

Cinnabar, or vermilion, was called by the Greeks *κιννάβαρι*,‡ and by the Romans minium. It is said by Theophrastus§ to have been discovered by Callias, an Athenian, ninety years before Praxibulus, and in the 349th year of Rome; and was prepared by washing the ores of quicksilver. According to Pliny,|| who quotes Verrius, it was a colour held in great esteem in Rome, at the time of the Republic; on great festivals, it was used for painting the face of Jupiter Capitolinus, and likewise for colouring the body of the victor in the triumphal processions; "Sic Camillum triumphasse."¶ Pliny mentions that even in his time vermilion was always placed at triumphal feasts amongst the precious ointments; and that the first occupation of new censors of the Capitol, was to fill the place of vermilion-painter to Jupiter.

Vermilion was always a very dear colour amongst the Romans; and we are informed by Pliny, that to prevent the price from being excessive, it was fixed by the government. The circumstance of the chambers of the baths of Titus being covered with it, affords proof in favour of their being intended for imperial use; and we

* Pliny, lib. xxxv. cap. 20.

† De Architectura, lib. vii. cap. 7.

‡ Dioscorides, lib. v. cap. 109.

§ De Lapid. cap. 104.

|| Lib. xxxiii. cap. 36.—Nunc inter pigmenta magnæ auctoritatis, et quondam apud Romanos non solum maximæ, sed etiam sacræ.

¶ Ibid.

are expressly informed by the author I have just quoted, that the Laocoon, in his time, was in the palace of Titus:* and the taste of the ancients, in selecting a colour to give full effect to their masterpieces of sculpture, was similar to that of a late celebrated English connoisseur.

Pliny describes a second, or inferior sort of vermilion, formed by calcining stone, found in veins of lead. It is evident that this substance was the same as our minium; and the Roman cerussa usta, and the stones alluded to by Pliny, must have been carbonate of lead: and he states distinctly, that it is a substance which becomes red only when burnt.

3. *Of the Yellows of the Ancients.*

A large earthen pot, found in one of the chambers of the baths of Titus, contains a quantity of a *yellow paint*, which, submitted to chemical examination, proved to be a mixture of yellow ochre with chalk or carbonate of lime.

This colour is used in considerable quantities in different parts of the baths, but principally in the least ornamented chambers, and in those which were probably intended for the use of the domestics. In the vase to which I alluded in the last section, I found three different yellows; two of them proved to be yellow ochre, mixed with different quantities of chalk; and the third a yellow ochre, mixed with red oxide of lead, or minium.

The ancients procured their yellow ochre† from different parts of the world; but the most esteemed, as we are informed by Pliny, was the Athenian ochre; and it

* Lib. xxxvi. cap. 4.—Sicut in Laocoonte, qui est in Titi Imperatoris domo, opus omnibus et picturæ et statuariæ artis præponendum.

† ὤχρα. Theophrastus de Lapidibus.

is stated by Vitruvius, that in his time the mine which produced this substance, was no longer worked.

The ancients had two other colours which were orange or yellow; the auripigmentum, or ἀρσενικον, said to approach to gold in its colour, and which is described by Vitruvius* as found native in Pontus, and which is evidently sulphuret of arsenic; and a *pale* sandarach, said by Pliny to have been found in gold and silver mines, and which was imitated at Rome by a partial calcination of ceruse, and which must have been massicot, or the yellow oxide of lead mixed with minium. That there was a colour called by the Romans sandarach, different from pure minium, is evident from what Pliny says—namely, that the palish kind of orpiment resembles sandarach, and from the line of Naevius, one of the most ancient Latin poets, "Merula sandaracino ore:" so that this colour must have been a bright yellow, similar to that of the beak of the blackbird.† Dioscorides describes the best, σανδαράχη, as approaching in colour to vermilion,‡ and the Greeks probably always applied this term to minium; but the Romans seemed to have used it in a different sense; and some confusion was natural, when different colours were prepared from the same substance, by different degrees of calcination.

I have not detected the use of orpiment in any of the ancient fresco paintings; but a deep yellow, approaching to orange, which covered a piece of stucco in the ruins near the monument of Caius Cestius, proved to be oxide of lead, and consisted of massicot, mixed with minium. It is probable that the ancients used many

* Vitruvius, lib. vii.

† Histoire de la Peinture ancienne, page 199.

‡ Lib. v. 122.

colours from lead of different tints between the usta of Pliny, which was our minium, and imperfectly decomposed ceruse, or pale massicot.

The yellows in the Aldovrandini picture, are all ochres. I examined the colours in a very spirited picture on the wall of one of the houses at Pompeii, of a lion and a man, they all proved to be red and yellow ochres.

4. *Of the Blue Colours of the Ancients.*

Different shades of blue are used in the different apartments of the baths of Titus, and several very fine blues exist in the mixtures of colours, to which I have referred in the last two sections.

These blues are pale or darker, according as they contain larger or smaller quantities of carbonate of lime, but when this carbonate of lime is dissolved by acids, they present the same body of colour,—a very fine blue powder similar to the best smalt or to ultramarine, rough to the touch, and which does not lose its colour by being heated to redness; but which becomes agglutinated and semifused at a white heat.

This blue I found was very little acted on by acids. Nitro-muriatic acid by being long boiled upon it gained, however, a slight tint of yellow, and afforded proofs of the presence of oxide of copper.

A quantity of the colour was fused for half an hour with twice its weight of hydrate of potassa; the mass, which was bluish-green, was treated by muriatic acid in the manner usually employed for the analysis of siliceous stones, when it afforded a quantity of silica equal to more than $\frac{3}{5}$ of its weight. The colouring matter readily dissolved in solution of ammonia, to which it gave a bright blue tint, and it proved to be oxide of copper. The residuum afforded a considerable quantity of alumine, and a small quantity of lime.

Amongst some rubbish that had been collected in one of the chambers of the baths of Titus, I found several large lumps of a deep blue frit, which, when powdered and mixed with chalk, produced colours exactly the same as those used in the baths, and which, when submitted to chemical tests, were found to be the same in composition.

The minute quantity of lime found in this substance was not sufficient to account for its fusibility; it was therefore reasonable to expect the presence of a fixed alkali in it; and on fusing some of it with three times its weight of boracic acid, and treating the mass with nitric acid and carbonate of ammonia, and afterwards distilling sulphuric acid from it, I procured from it sulphate of soda, which proves that it was a frit made by means of soda, and coloured with oxide of copper.

The undiluted colour in its form of frit is used for ornamenting some of the mouldings detached from the ceilings of the chambers in the baths of Titus; and the walls of one chamber between the compartments of red marble bear proofs of having been covered with this frit, and contain a considerable quantity of it.

There is every reason to believe that this is the colour described by Theophrastus as discovered by an Egyptian king,* and of which the manufacture is said to have been anciently established at Alexandria.

Vitruvius speaks of it under the name of cæruleum,† as the colour used commonly in painting chambers, and states, that it was made in his time at Puzzuoli, where the method of fabricating it was brought from Egypt by Vestorius; he gives the method of preparing it by

* De Lapidibus, sect. xcviii.

† Lib. vii. cap. 11.

heating strongly together sand, flos nitri,* and filings of copper.

Pliny mentions other blues, which he calls species of sand (arenæ) from the mines of Egypt, Scythia, and Cyprus. These natural blues, there is reason to believe, were different preparations of lapis lazuli, and of the blue carbonates and arseniates of copper.

Both Pliny and Vitruvius speak of the Indian blue, which the first author states to be combustible, and which was evidently a species of indigo.

I have examined several blues in the fragments of fresco painting from the ruins near the monument of Caius Cestius. In a deep blue approaching in tint to indigo, I found a little carbonate of copper, but the basis of the colour was the frit before described.

The blues in the Nozze Aldovrandine, from thus resisting the action of acids, and from the effects of fire, I am inclined to consider as composed of the Alexandrian or Puzzuoli blue.

In an excavation made at Pompeii, in May, 1814, at which I was present, a small pot containing a pale blue colour was dug up, which the exalted personage† by whose command the excavation was made, was so good as to put into my hands. It proved to be a mixture of carbonate of lime with the Alexandrian frit.‡

Vitruvius states, that the ancients had a mode of

* This identifies the nitrum of the ancients with carbonate of soda.

† [Murat, then king, from whom the author experienced much courtesy;—a patron and admirer of the fine arts,—more was done during his short and feverish reign in bringing to light the treasures of antiquity buried in Pompeii than before or since in a triple space of time.]

‡ This probably is the same colour as that examined by M. Chaptal. He did not search in it for alkali, or there is every reason to suppose he would have found soda.

imitating the Indian blue or indigo, by mixing the powder of the glass called by the Greeks ὕαλος, with selinusian "creta" or annularian "creta," which was white clay or chalk mixed with stained glass; the same practice is likewise referred to by Pliny.

There is much reason for supposing that this stained glass, or ὕαλος, was tinged with oxide of cobalt; and that these colours were similar to our smalt. I have not found any powdered colour of this kind in the baths of Titus, or in any other Roman ruins; but a blue glass tinged with cobalt is very common in these ruins, which when powdered forms a pale smalt.

I have examined many pastes and glasses that contain oxide of copper, they are all bluish-green, green, or of an opaque watery blue. The transparent blue glass vessels which are found with vases in Magna Græcia are tinged with cobalt, and on analyzing different ancient transparent blue glasses which Mr. Millingen was so good as to give me, I found cobalt in all of them.*

Theophrastus, in speaking of the manufacture of glass, states as a report that "χαλκὸς" was used to give it a fine colour, and it is extremely probable that the Greeks took cobalt for a species of χαλκὸς. I have examined some Egyptian pastes which are all tinged blue and green with copper; but though I have made experiments on nine different specimens of ancient Greek and Roman *transparent* blue glass, I have not found copper in any, but cobalt in all of them.†

* The mere fusion of these glasses with alkali, and digestion of the product with muriatic acid, was sufficient to produce a sympathetic ink from them; even the silica separated by the acid gained a faint blue-green tint by heat, and the solution in muriatic acid became permanently green by the action of sulphuric acid, a phenomenon, Dr. Marcet has observed as belonging to muriate of cobalt.

† A gentleman, at Milan, informed me last summer, that he had

5. *Of the ancient Greens.*

The ceiling of the chambers called the Baths of Livia, is highly ornamented with gildings and paintings; the larger paintings have been removed, but the ground-work and the borders remain. A fragment detached from the borders, which appears of the same colour as the ground-work was of a deep sea-green. The colouring matter examined, proved to be soluble in acids with effervescence, and when precipitated from acids, it re-dissolved in solution of ammonia, giving it the bright blue tint produced by oxide of copper. There are several different shades of green employed in the baths of Titus, and on the fragments found near the monument of Caius Cestius; in the vase of mixed colours I found three different varieties; one, which approached to olive, was the common green earth of Verona; another, which was pale grass-green, had the character of carbonate of copper mixed with chalk; and a third, which was sea-green, was a green combination of copper mixed with the blue copper frit.

All the greens that I examined on the walls of the baths of Titus, were combinations of copper. From the extreme brilliancy of a green which I found in the vineyard to which I have so often referred, I suspected that it might contain arsenious acid, and be analogous to Scheele's green; but on submitting it to experiments, it afforded no indications of this substance, and proved to be a pure carbonate of copper.

found the oxide of cobalt in the blue glass found in the ruins of Hadrian's villa, and at that time I had no idea that cobalt was known to the ancients. Mr. Hatchett and Mr. Klaproth had both found oxide of copper in some ancient blue glasses, which I conceive must have been opaque.

The greens of copper were well known to the Greeks; the most esteemed is described by Theophrastus and Dioscorides under the name of χρυσοκόλλα, and is stated by both to be found in metallic veins.

Vitruvius mentions chrysocolla as a native substance found in copper mines, and Pliny speaks of an artificial chrysocolla made from the clay found in the neighbourhood of metallic veins, which clay was most probably impregnated with copper. He describes it as rendered green by the herb luteum. There is every reason to believe, that the native chrysocolla was carbonate of copper, and that the artificial was clay impregnated with sulphate of copper rendered green by a yellow dye.

Some commentators have supposed that chrysocolla is the same substance as borax, because Pliny has mentioned that a preparation called by this name was used by goldsmiths for soldering gold*; but nothing can be more gross than this mistake, which, however, has been copied into many elementary books of chemistry. The material used for soldering gold consisted of carbonate or oxide of copper mixed with alkaline phosphates. This is evident from the description of Dioscorides, "Περὶ ιοῦ σκώληκος," lib. v. c. 92, who says it was prepared from urine treated in brass mortars. Pliny says likewise, that it was prepared from "Cypria ærugine et pueri impubis urina addito nitro."† The name of chrysocolla was probably derived from the green powder used by the goldsmiths, and which contained carbonate of copper as one of its ingredients.‡

* Hist. de la Peinture ancienne, p. 38, "Nos droguistes la nomme Borax."

† Lib. xxxiii. cap. 5.

‡ The commentators have been likewise misled by Pliny's description, "chrysocolla humor est in puteis per venam auri defluens, &c." ibid; but this is merely an inaccurate account of the decomposition of

Amongst the substances found in the baths of Titus were some masses of a grass-green colour. I at first thought these might be specimens of native chrysocolla; they proved indeed to be carbonate of copper, but it had formed round longitudinal nuclei of red oxide of copper, so that probably these substances had been copper nails or small pieces of copper used in the building, converted by the action of the air, during so many centuries, into oxide and carbonate.

The ancients, as it appears from Theophrastus, were well acquainted with verdigrise. Vitruvius mentions it amongst pigments, and probably many of the ancient greens, which are now carbonate of copper, were originally laid on in the state of acetite.

The ancients had beautiful deep-green glasses, which I find are tinged with oxide of copper; but it does not appear that they used their glasses in a state of powder as pigments.

The greens in the Aldovrandini picture are all of copper, as was evident from the action of muriatic acid upon them.

6. *Of the Purple of the Ancients.*

The Πορφῦρα of the Greeks, and the ostrum of the Romans was regarded as their most beautiful colour, and was prepared from shell-fish.

Vitruvius* says that the colour differed according to the country from which the shell-fish was brought: that it afforded a colour deeper and more approaching to

a vein containing copper. We have no reason for supposing that the Greeks and Romans were acquainted with borax. Pliny, probably misled by the application of the same name to different substances, considered chrysocolla as the cement of gold in mineral veins.

* Lib. vii. cap. 13.

violet from the northern countries, and a redder colour from the southern coasts. He states that it was prepared by breaking the fish with instruments of iron, freeing the purple liquor from the shell containing it, and mixing it with a little honey: and Pliny says, that for the use of the painters argentine "creta"* was dyed with it; and both Vitruvius and Pliny say that it was adulterated, or imitations of it made, by tinging "creta" with madder† and "hysginum." The finest purple, Pliny says, had a tint like that of a deep-coloured rose; and in painting, he states that it was laid on to give the last lustre to the sandyx, a composition made by calcining together red ochre and sandarach, and which, therefore, must have been nearly the same as our crimson.

In the baths of Titus there is a broken vase of earthenware, which contains a pale rose colour; where it has been exposed to air, it has lost its tint, and is become of a cream colour, but the interior has a lustre approaching to that of carmine.

I have made many experiments on this colour. It is destroyed and becomes of a red brown by the action of concentrated acids and alkalies; but diluted acids dissolve a considerable quantity of carbonate of lime with which the body colour is mixed, and leave a substance

* Probably a clay used for polishing silver. The ancients were not acquainted with the distinction between aluminous and calcareous earths, and *creta* was a term applied to every white fine earthy powder.

† Madder was extensively used by the ancients in dying, and from this passage it is probable that they were acquainted with the art of making a lake from it similar to that used by modern painters. It was probably one of the colours used by the Egyptians in dying their stuffs of different colours from the same liquor, by means of mordants. If we can trust Pliny's account, they practised calico printing in a manner similar to the moderns. Lib. xxxv. cap. 42.

of a bright rose colour; this substance, when heated, first blackens, and when urged with a strong flame, becomes white; and treated with alkali, proves to be composed of siliceous aluminous and calcareous earths, with no sensible quantity of any metallic substance, except oxide of iron.

I endeavoured to discover if the colouring matter was combustible. It was gradually heated on a glass tube filled with oxygen; it did not inflame, but became red-hot sooner than it would have done had it been merely earthy matter: on exposing the gas in the tube to lime-water, there was a precipitation of carbonate of lime. Some of it was mixed with hyperoxymuriate of potassa, and heated in a small retort; when the salt fused, there was a slight scintillation, a little moisture appeared, and the gas given off received into lime water, occasioned a very evident precipitation.

It appeared from these experiments that the colouring matter was a compound of either vegetable or animal origin. I threw some of it apon a hot iron: it emitted scarcely any smoke, and gave a smell which had some resemblance to that of prussic acid, but which was extremely faint.

When hydrate of potassa was fused in contact with it, the vapours that rose had no distinct ammoniacal smell; they gave indeed slight fumes to paper moistened with muriatic acid, but this is far from being an unequivocal proof of animal matter. I compared this colour with vegetable lake from madder, and animal lake from cochineal diluted to the same degree as nearly as could be judged, and fixed upon clays. The lake of madder, after being dissolved in strong muriatic acid, had its colour restored by alkalies, which was not the case with the ancient lake. The lake of madder likewise gave a much deeper tint to muriatic acid, and produced a tawny hue when its weak muriatic solution was acted on by muriate of iron, whereas the ancient lake did not change in colour. The ancient lake agreed with the lake of cochineal in being rendered of a deeper hue by weak alkalies, and of a brighter hue by weak acids; but it differed from it in being much more easily destroyed by strong acids. It agreed with both the vegetable and animal lakes in being immediately destroyed by a solution of chlorine.

The lake made from cochineal produced much denser fumes when exposed to fused potash, and afforded a distinct ammoniacal smell. The two modern lakes when burnt in oxygen did not give stronger signs of inflammation then the ancient. I ascertained the loss of weight this ancient lake suffered by combustion, and found it only $\frac{1}{30}$, and this loss must in great part have depended on the expulsion of water from the clay on which it was fixed. This circumstance induced me to renounce the idea of attempting to determine its nature from the products of its decomposition, which, in the case of so small a quantity of matter diffused over so large a quantity of surface, could not have afforded unequivocal results.

The durability of this lake, whether vegetable or animal, is a very curious circumstance; but the exterior part which has been exposed to air has suffered. This durability probably depends in a great measure upon the attractive powers of so large a mass of alumina; for whenever one proportion of a substance is combined with many proportions of another substance, it is very difficult to decompose or detach the one proportion.

From the circumstances which have been noticed

respecting this colour, it is impossible to form an opinion whether it is of vegetable or animal origin. If of animal origin, it is most probably the Tyrian or marine purple; and by some comparative experiments on the purple obtained from shell-fish, the question might perhaps be decided.* It is very probable that the most expensive colour would be employed for ornamenting the imperial baths; and it is not impossible that Pliny may have alluded to the palace of the Cæsars when he says "nunc et purpuris in parietes mĭgrantĭbus, et Īndĭa confĕrente flŭmĭnum suorum limum, et draconum et elephantorum saniem, nulla nobilis pictura est." Lib. xxxv. cap. 32.

I have seen no colours of the same tint as this ancient lake in any of the ancient paintings in fresco. The purplish reds in the baths of Titus are mixtures of red ochres, and the blues of copper. In the Aldovrandini picture there is a purple in the garment of the Pronuba, but of an inferior hue; and this purple appears to be a compound mineral colour of the nature of these. It was not destroyed by solution of chlorine; and when a little of it was exposed to muriatic acid, it rendered the acid yellow, and the remainder yielded a residual blue powder.

* M. Chaptal considers the lake he found amongst the colours from Pompeii (as I have already mentioned) as of vegetable origin; and he founds his opinion upon the circumstance of its not affording by decomposition the smell peculiar to animal substances; but probably this smell, even if produced by recent purple colouring matter of animal origin, would not belong to colouring matter of 1700 years old. For it is most probably owing merely to albumen or gelatine not essential to the colouring particles, and much more rapidly decomposed.

7. *Of the Blacks and Browns of the Ancients.*

There is one chamber in the baths of Titus, of which the ground-work is black. I have found several fragments of stucco painted black, both in the baths of Titus and in the vineyard above mentioned, and also in some ruins near the Porta del Popolo. I scraped off some of these colours, and submitted them to experiments; they were not acted on by acids or alkalies, they deflagrated with nitre, and had all the properties of pure carbonaceous matter.

I found no blacks, but three different shades of brown in the vase of mixed colours; one was snuff colour, one deep red brown, and the third a dark olive brown. The two first proved to be ochres, which had been probably partially calcined; the third contained oxide of manganese, as well as oxide of iron, and afforded chlorine when acted on by muriatic acid.

All the ancient authors describe the artificial Greek and Roman blacks as carbonaceous, and made either from the powder of charcoal or the decomposition of resin, (a species of lamp-black,) or from the lees of wine, or from the common soot of wood fires. Pliny mentions the inks of the cuttle-fish, but says, "ex his non fit."* Some years ago I examined this substance, and found it a carbonaceous body mixed with gelatine. Pliny speaks of ivory black as invented by Apelles; he says, likewise, that there is a natural fossil black, and another black prepared from an earth of the colour of sulphur. Probably both these substances are ores of iron and manganese.

That the ancients were acquainted with the ores of

* i. e. the atramentum.

manganese is evident from the use made of it in colouring glass. I have examined two specimens of ancient Roman purple glass, both of which were tinged with oxide of manganese. Pliny speaks of different brown ochres, and particularly of one from Africa, which he names Cicerculm, which probably contained manganese: and Theophrastus mentions a fossil* which inflamed when oil was poured upon it, a property belonging to no other fossil substance now known but the *black wad*, an ore of manganese, and which is now found in Derbyshire.

The browns in the paintings of the baths of Livia, and in the Aldovrandini picture, are all produced by mixtures of ochres with blacks. Those in the Aldovrandini picture yield oxide of iron to muriatic acid, but the darker shades were not touched by that acid, nor by solution of alkalies.

8. *Of the Whites of the Ancients.*

The white colours in the Aldovrandini picture are soluble in acids with effervescence, and have the characters of carbonate of lime.

The principal white in the vase of mixed colours appears to be a very fine chalk. There is another white with a tint of cream colour, which is a fine aluminous clay.

The whites that I have examined from the baths of Titus, and those from other ruins, are all of the same kind.

I have not met with ceruse amongst the ancient colours, though we know from Theophrastus, Vitruvius,

* Theophrastus says it is like decomposed wood—*παρόμοιος ὢν ξύλῳ σαπρῷ*, 12th page of John de Laet's edition.

and Pliny, that it was a common colour: and Vitruvius describes it as made by the action of lead upon vinegar.

Several white clays are mentioned by Pliny as employed in painting, of which the Paraetonium was considered as affording the finest colour.

9. *Of the Manner in which the Ancients applied their Colours.*

It appears from Vitruvius that the colours used in fresco painting were applied moist to the surface of stucco* formed of powdered marble cemented by lime; he states that the wall or ceiling had three distinct coatings of stucco made of this material, of which the first contained coarse powder of marble, the second the finer powder, and the third the finest powder of all, and that after this the wall was polished before the colour was applied. The stuccos that remain in the ruins of the baths of Titus and Livia are of this kind, and so is the ground of the Aldovrandini picture; they are beautifully white, and almost as hard as marble, and the granular marble of different degrees of fineness may be distinguished in them. This circumstance, indeed, offers a test of the antiquity of ruins at Rome. In the houses that have been built in the middle and later ages, decomposing lava has been mixed with the calcareous cement instead of granular marble, and the stuccos of these houses are grey or brown, and very coarse in their texture.

Pliny says that purple, orpiment, ceruse, the natural azure, indigo, and the meline white, were injured by application to wet stucco, which is easily explained in the case of orpiment, carbonate of copper, ceruse, and indigo, from their chemical composition.

* Lib. vii. cap. 2, 3, and 4.

Vitruvius states that in fresco painting vermilion changed if exposed to light, and he recommends the encaustic process for fining the colour under this circumstance; namely, laying over it a coat of punic wax, and liquifying the wax so as to make a varnish for the colour.

Pliny describes this process as applied in painting ships; and we know from his authority that several pictures of the great Greek masters were painted in encaustic, and that the different colours were laid on mixed with wax. I have examined several pieces of the painted stuccos found in the different ruins, and likewise the Aldovrandini picture, with a view of ascertaining if any application had been made to fix the colour; but neither by the test of alcohol, nor by heat, nor by the action of water, could I detect the presence of any wax varnish, or animal or vegetable gluten.

The pot of colours to which I have already referred, found at Pompeii, was blackened by smoke, as if it had recently been on a fire of wood. I thought that this might be owing to some process of dissolving gluten or varnish in the preparation of the colour; but I could detect no substance of this kind mixed with the colouring matter.

Pliny states, that gluten (our glue)* was used in painting with blacks; and this specific mention of its application would induce the belief that it was not employed with other colours, which adhered without difficulty to, and were imbibed by a surface so polished and well prepared as the Roman stucco; and the lightness of carbonaceous matter alone probably rendered this application necessary.

* Lib. xxxv. cap. 25. "Omne atramentum sole perficitur, librarium gummi tectorium glutino admixto."

10. *Some General Observations.*

It appears from the facts that have been stated, and the authorities quoted, that the Greek and Roman painters had almost the same colours as those employed by the great Italian masters at the period of the revival of the arts in Italy. They had, indeed, the advantage over them in two colours, the Vestorian or Egyptian azure, and the Tyrian or marine purple.

The azure, of which the excellence is proved by its duration for seventeen hundred years, may be easily and cheaply made; I find that fifteen parts by weight of carbonate of soda, twenty parts of powdered opaque flints, and three parts of copper filings strongly heated together for two hours, gave a substance of exactly the same tint, and of nearly the same degree of fusibility, and which, when powdered, produced a fine deep sky-blue.

The azure, the red and yellow ochres, and the blacks are the colours that seem not to have changed at all, in the ancient fresco paintings. The vermilion is darker than recently-made Dutch cinnabar, and the red lead is inferior in tint to that sold in the shops. The greens in general are dull.

The principle of the composition of the Alexandrian frit is perfect; namely, that of embodying the colour in a composition resembling stone, so as to prevent the escape of elastic matter from it, or the decomposing action of the elements; this is a species of artificial lapis lazuli, the colouring matter of which is naturally inherent in a hard siliceous stone.

It is probable that other coloured frits may be made, and it is worth trying whether the beautiful purple colour given by oxide of gold, cannot be made useful in painting in a densely tinted glass.

When frits cannot be employed, metallic combinations which are insoluble in water, and which are saturated with oxygen, or some acid matter, it is evident from the proof of a duration of seventeen centuries, are the best pigments. In the red ochres the oxide of iron is fully combined with oxygen and carbonic acid; and these colours have not changed. The carbonate of copper which contains an oxide and an acid have changed very little.

Massicot and orpiment were probably the least permanent amongst the ancient mineral colours. Of the colours, the discovery of which is owing to the improvements in modern chemistry, the patent yellow is much more durable than any ancient yellow of the same brilliancy; and chromate of lead, an insoluble compound of a metallic acid with a metallic oxide, is a much more beautiful yellow than any possessed by the ancients, and there is every reason to believe, is quite unalterable.

Scheele's green (the arsenite of copper), and the insoluble muriatic combination of copper, will probably be found more unalterable than the ancient greens; and the sulphate of baryta offers a white superior to any possessed by the Greeks and Romans.

I have tried the effect of light and air upon some of the colours formed by the new substance iodine. Its combination with mercury offers a good red; but it is, I think, less beautiful than vermilion, and it appears to change more by the action of light.

Its compound with lead gives a beautiful yellow, little inferior to the chromate of lead; and I possess some of this colour which has been exposed to light and air without alteration for several months.

In many of the figures and ornaments in the outer chambers of the baths of Titus, where only outlines or spots remain, or shades of ochre, it is probable that vegetable or animal colours, such as indigo and the different dyed clays, were used.*

Pliny speaks of the celebrated Greek painters as employing only four colours. "Quatuor coloribus solis immortalia illa opera facere: ex albis Melino, ex silaceis Attico, ex rubris Sinopide Pontica, ex nigris atramento, Apelles, Echion, Melanthius, Nicomachus, clarissimi pictores;"† but as far as Apelles and Nicomachus are concerned, this is a mistake: and it is not unlikely that Pliny was misled by an imperfect recollection of a passage in Cicero, who describes the earlier Greek school as using only four colours; but the later Greek painters as perfect masters in all the resources of colouring. "Similis in pictura ratio est: in qua Zeuxim et Polygnotum, et Timantem, et eorum, qui non sunt usi plus quam quatuor coloribus, formas et lineamenta laudamus: at in Aetione, Nicomacho, Protogene, Apelle, jam perfecta sunt omnia." Cicero, Brutus, seu de claris oratoribus, c. 18. Pliny himself describes with enthusiasm the Venus ἀναδυομένη of Apelles: and in this picture the sea was represented, which required azure.

The Greek painters, like the most illustrious artists of the Roman and Venetian school, were probably,

* Some excellent pictures have suffered very much in modern times from the same cause; the lakes in the frescos of the Vatican have lost much of their brilliancy, which they must have possessed originally. The blues in many pictures of Paul Veronese are become muddy.

† Lib. xxxv. c. 32.

however, sparing in the use of the more florid tints in historical and moral painting, and produced their effects rather by the contrasts of colouring in those parts of the picture where a deep and uniform tint might be used, than by brilliant drapery.

If red and yellow ochres, and blacks and whites, were the colours most employed by Protogenes and Apelles, so they are likewise the colours most employed by Raphael and Titian in their best style. The St. John and the Venus, in the tribune of the gallery at Florence, offer striking examples of pictures in which all the deeper tints are evidently produced by red and yellow ochres, and carbonaceous substances.

As far as colours are concerned, these works are prepared for that immortality which they deserve; but unfortunately the oil and the canvass are vegetable materials, and liable to decomposition, and the last less durable than even the wood on which the Greek artists painted their celebrated pictures.

It is unfortunate that the materials for receiving these works, which are worthy of passing down to posterity as eternal monuments of genius, taste, and industry, are not imperishable marble,* or stone: and that frits, or unalterable metallic combinations, have not been the only pigments employed by great artists; and that their varnishes have not been sought for amongst the transparent combinations of the earths with water, or

* Copper, it is evident, from the specimens in the ruins of Pompeii, is a very perishable material, and, therefore, even enamels made on copper will yield to time. Canvass, by being impregnated with bitumen, is rendered much more durable, as is evident from the duration of the linen impregnated with bitumen and asphaltum used for infolding the Egyptian mummies.

amongst the crystalline compounds unalterable in the atmosphere.*

* The artificial hydrat of alumina will probably be found to be a substance of this kind: possibly the solution of boracic acid in alcohol will form a varnish. The solution of sulphur in alcohol is likewise worthy of an experiment. Many other similar combinations might be named.

Rome, Jan. 14*th*, 1815.

III.

SOME OBSERVATIONS AND EXPERIMENTS ON THE PAPYRI FOUND IN THE RUINS OF HERCULANEUM.*

In a paper† intended for private circulation only on the MSS. found in the excavations made at Herculaneum, but which was published by mistake in the Journal of Science and the Arts, I have described in a general manner the circumstances which led me to make experiments on these remains, and I mentioned some of my first observations on this subject. Mr. Hamilton, to whom this communication was sent, entered into my views with all that ardour for promoting the progress of useful knowledge which so peculiarly belongs to his character, and on his representation of them, the Earl of Liverpool and Viscount Castlereagh, with the greatest liberality, placed at my disposal such funds as were requisite for paying the persons whom it was necessary to employ in trying new chemical methods of unrolling the MSS., and for examining and preserving them when unrolled; and his present Majesty, then Prince Regent, graciously condescended to patronize the undertaking.

* [From the Phil. Trans. for 1821. Read before the Royal Society, March 15th, 1821.]

† [Dated Rome, Feb. 12, 1819; published in vol. vii. of the Journal above referred to. As the particulars of most interest in that paper are to be found in this subsequent and more finished one, it has not appeared advisable to insert the former.]

In this communication, I shall do myself the honour of laying before the Royal Society an account of all that I have been able to do on this subject; namely, first, a detail of my early experiments in England on fragments of papyri, which induced me to believe that chemistry might afford considerable assistance towards unrolling the MSS. Secondly, a description of the rolls in the museum at Naples, and of some analytical experiments I made upon them. Thirdly, a detail of the various chemical processes carried on in the museum at Naples on the MSS., and of the reasons which induced me to renounce my undertaking before it was completed. And lastly, some general observations on the MSS. of the ancients.

I trust these matters will not be found wholly devoid of interest by the Society, and that they will excuse some repetitions of what I have stated in the report before referred to, as they are necessary for a complete elucidation of the subject.

1. *An Account of some Experiments made in England on Fragments of Papyri, in* 1818.

In examining, chemically, some fragments of a roll of papyrus found at Herculaneum, the leaves of which adhered very strongly together, I found that it afforded, by exposure to heat, a considerable quantity of gaseous matter, which was principally inflammable gas, and when acted on by muriatic or nitric ether, it coloured them; and when it was exposed to heat after the action of these fluids, there was an evident separation of the leaves of the MSS.

Chlorine and iodine, it is well known, have no action upon pure carbonaceous substances, and a strong attraction for hydrogen; and it occurred to me, that these bodies might with propriety be used in attempting to destroy the matter which caused the adhesion of the leaves, without the possibility of injuring the letters on the papyri, the ink of the ancients, as is well known, being composed of charcoal.

Having through the polite assistance of Sir Thomas Tyrwhitt procured some fragments of papyri on which Dr. Sickler, and some on which Dr. Hayter had operated, and by the kindness of Dr. Young a small portion of an MS. which he had himself unsuccessfully tried to unroll, I made some experiments upon them, by exposing them to the action of chlorine and the vapour of iodine, heating them gently after the process. These trials all afforded more or less hopes of success. When a fragment of a brown MS. in which the layers were strongly adherent, was placed in an atmosphere of chlorine, there was an immediate action; the papyrus smoked and became yellow, and the letters appeared much more distinct; and by the application of heat the layers separated from each other, giving off fumes of muriatic acid. The vapour of iodine had a less distinct action, but still a sensible one; and it was found that by applying heat alone to a fragment in a close vessel filled with carbonic acid or the vapour of ether, so as to raise the heat very gradually, and as gradually to lower it, there was a marked improvement in its texture, and it was much more easily unrolled.

Even in these preliminary trials I found that it was necessary to employ only a limited and small quantity of chlorine, too large a quantity injuring the texture of the layer, and decomposing the earths which it contained; and that the action of heat was much more efficaceous when the MS. had previously been exposed to chlorine, as the muriatic acid vapour formed greatly assisted the separation of the leaves, and a smaller degree of heat was required. But in all the trials, I found the success absolutely depended upon the manner in which the temperature was regulated. When the fragment was too rapidly heated, the elastic fluid disengaged usually burst the folds of the MS.; and when the heat was lowered too suddenly, the layers sometimes split in irregular parts, probably from the sudden contraction consequent on quick cooling.

From the products of the distillation of these fragments, which were water, acetous acid, ammonia, carbonic acid, and much inflammable gas, I inferred, that the papyri to which they belonged must contain much undecomposed vegetable matter, and could not be purely carbonaceous; but as there were great differences in the appearances even of the few papyri in England, which had been presented to his Majesty George IV., when Prince of Wales, an opinion on this subject was more likely to be correct when formed after an examination, not only of all the MSS. found at Herculaneum, but likewise of the circumstances of the excavations made there; and I had an opportunity, during the time I remained at Naples, in two successive winters,* to satisfy my mind on this subject, and to obtain the information which will be given in the next section.

2. *On the State of the MSS. found at Herculaneum.*

The persons who have the care of the MSS. found at Herculaneum, state that their original number was 1696, and that 431 have been operated upon or pre-

* [Part of the winters of 1819, and 1820. Vide Vol. I. pp. 237, 255.]

sented to foreign governments, so that 1265 ought to remain; but amongst these, by far the larger proportion are small fragments, or specimens so injured and mutilated, that there is not the least chance of recovering any portion of their contents; and when I first examined the rolls in detail in January 1819, it did not appear to me, that more than from 80 to 120 offered proper subjects for experiments; and this estimate, as my researches proceeded, appeared much too high. These MSS. had been objects of interest for nearly seventy years;* the best had long ago been operated upon, and those remaining had not only undergone injuries from time, but likewise from other causes, such

* [The reader may be reminded that the catastrophe of the overwhelming of Herculaneum took place in A.D. 79; that in modern times (for there is reason to infer it was partially explored before) the ruins were discovered accidentally in sinking a well in 1713; but that no progress was made in excavating them for many years after. In the Philosophical Transactions, for 1740, is the following notice of the event, the earliest which occurs in that collection, written by W. Sloane, Esq., entitled, "Discovery of the Remains of a City underground near Naples."

"At Resina, about four miles from Naples, under the mountain, within half a mile of the sea-side, there is a well, down which about thirty yards is a hole, which some people have the curiosity to creep into, and may afterwards creep a good way under ground, and with lights find foundations of houses and streets, which, by some it is said, was in the time of Romans called Aretina, others say Port Hercules, where the Romans usually embarked for Africa. Mr. S. has seen the well, which is deep, and has a good depth of water at the bottom, but he never cared to venture down, being heavy and the ropes bad. This city, it is thought, was overwhelmed by an irruption of the mountain Vesuvius, not sunk by earthquake, as were Cuma, Baia, Trepergola, &c." Phil. Trans. Abridg. vol. viii. p. 403. And, in another later notice in the same collection, Sig. Camillo Paderni, writing from Naples, April 27, 1754, states, "In one of the buildings there has been found an entire library composed of volumes of Egyptian papyrus, of which 250 have been taken out, and the place is not yet cleared or emptied." Phil. Trans. Abridg. vol. x. p. 494.]

as transport, rude examination, and mutilations for the purpose of determining if they contained characters.

The appearances of the different rolls were extremely various. They were of all shades of colours from a chestnut brown to a deep black; some internally were of a glossy black, like jet, which the superintendents called "varnished;" several contained the umbilicus or rolling-stick in the middle converted into dense charcoal. I saw two or three specimens of papyri which had the remains of characters on both sides, but in general one side only was written upon. In their texture they were as various as in their colours; the pale brown ones in general presented only a kind of skeleton of a leaf in which the earthy matter was nearly in as large a proportion as the vegetable matter, and they were light, and the layers easily separated from each other. A number of darker brown ones, which, from a few characters discovered in opening them, appeared to be Latin MSS., were agglutinated, as it were, into one mass; and when they were opened by introducing a needle between the layers, spots or lines of charcoal appeared where the folds had been, as if the letters had been washed over by water, and the matter of which they were composed deposited on the folds. Amongst the black MSS. a very few fragments presented leaves which separated from each other with considerable facility, and such had been for the most part operated upon; but in general the MSS. of this class were hard, heavy, and coherent, and contained fine volcanic dust within their folds. Some few of the black and darker brown MSS., which were loose in their texture, were almost entirely decayed, and exhibited on their surface a quantity of brown powder.

The persons to whom the care of these MSS. is confided, or who have worked upon them, have always attributed these different appearances to the action of fire, more or less intense, according to the proximity of the lava, which has been imagined to have covered the part of the city in which they were found; but this idea is entirely erroneous, that part of Herculaneum being, as I satisfied myself by repeated examinations, under a bed of tufa formed of sand, volcanic ashes, stones, and dust, cemented by the operation of water (probably at the time of its action in a boiling state).* And there is great reason to conclude, that the different states of the MSS. depend upon a gradual process of decomposition: the loose chestnut ones probably not having been wetted, but merely changed by the reaction of their elements, assisted by the operation of a small quantity of air; the black ones, which easily unroll, probably remained in a moist state without any percolation of water; and the dense ones, containing earthy matter, had probably been acted on by warm water, which not only carried into the folds earthy matter suspended in it, but likewise dissolved the starch and gluten used in preparing the papyrus and the glue of the

* [In confirmation, an extract may be given from a paper in the Phil. Trans. for 1753, by Signor Camillo Paderni, "On the Antiquities dug up from the Ancient Herculaneum." "The things of which Sig. Paderni says he had the charge, are many and extraordinary, consisting of metals; that is, bronzes, silver and gold of all kinds, of excellent workmanship. Beautiful cameos and intaglios. Glass of all sorts. Various productions of the earth; such as grain, beans, figs, dates, nuts, pistachios, almonds, rice, bread. Colours for painting. Medicines in pills, and other forms, with their marks. A phial of oil. Gold lace, perfectly well preserved, and extremely curious on account of its being made with massy gold, spun out, without any silk or other yarn. Soap, bran, and a variety of other things which it were tedious here to enumerate. There were found many volumes of papyrus, but turned to a sort of charcoal, and so brittle, that, being touched, it fell to ashes." Phil. Trans. Abridg. vol. x. p. 328.]

ink, and distributed them through the substance of the MSS., and some of these rolls had probably been strongly compressed when moist in different positions.

The operation of fire is not at all necessary for producing such an imperfect carbonization of vegetable matter as that displayed by the MSS. Thus, at Pompeii, which was covered by a shower of ashes, that must have been cold, as they fell at a distance of seven or eight miles from the crater of Vesuvius,* the wood of the houses is uniformly converted into charcoal; yet the colours on the walls, most of which would have been destroyed or altered by heat, are perfectly fresh, and where papyri have been found in these houses, they have appeared in the form of white ashes, as if of burnt paper; an effect produced by the slow action of the air penetrating through the loose ashes, and which had been impeded or prevented in Herculaneum by the tufa, which, as it were, has hermetically sealed up the town, and prevented any decay, except such as occurs in the spontaneous decomposition of vegetable substances, exposed to the limited operation of water and air; for instance, peat and Bovey coal.

The results of the action of heat upon the different specimens of the papyri, proved likewise, that they had never before been exposed to any considerable degree of temperature.

* [In confirmation of the above remark, I may refer to the observations I had an opportunity of making on the temperature of the air and water in the neighbourhood of the volcano which occurred in the Mediterranean in 1831; both were found actually cooler than the atmosphere and sea uninfluenced by the eruption. This probably was partly owing to the cooling effect of evaporation on the water ejected, and partly to that of elevation on the ashes and cinders, which, falling from the great height to which they had been projected, brought down a lower temperature than that of the surface.—Vide Phil. Trans. for 1832.]

Various specimens of papyri were heated to dull redness in a small covered crucible of platinum to which air had no access. Some of the chestnut and most perfect specimens lost nearly half their weight, and the very black ones, and those containing the largest quantity of white ashes, all lost more than one-third, as the following results, selected from a number, will show:—

No. 1. 100 parts of a pale chestnut papyrus lost 45 parts.

No. 2. 100 parts of a decomposed papyrus, chestnut coloured, but darker, lost 43.

No. 3. 100 parts of a very black papyrus lost 42.

No. 4. 100 parts of a pale papyrus, extremely loose in texture and partly converted into white ashes, lost 41.

No. 5. 100 parts of another of the same kind, lost 38.

When the whole of the carbonaceous and vegetable matter of the papyrus was destroyed by slow combustion, the white ashes remaining, which were principally carbonate of lime and lime, proved to be from $\frac{1}{10}$ to $\frac{1}{20}$ of the original weight of the papyrus; and in those specimens which were most dense, and that contained a white powder, the proportion of ashes was greater, and a larger quantity was insoluble in acids.

Ammonia was found in the products of all the papyri that I distilled, but least in those which contained no distinct characters; from which it is probable that it arose principally from decomposed glue, used in the manufacture of the ink, and which had been principally dissolved and carried off in those papyri which had been most exposed to the action of water.

I ascertained that what the Neapolitans called varnish, was decomposed skin, that had been used to infold some of the papyri, and which by chemical changes had produced a brilliant animal carbonaceous substance; this substance afforded abundance of ammonia by distillation, and left ashes containing much phosphate of lime.

3. *An Account of the Experiments on Papyri, made in the Museum at Naples.*

Only one method, and that a simple mechanical one, has been adopted for unrolling the MSS. It was invented by Padre Piazzi, a Roman, and consists in attaching thin animal membrane by a solution of glue to the back of the MSS., and carefully elevating the layers by silk threads when the glue is dry.

In considering this method in its general application, some circumstances occurred to me which afforded an immediate improvement. A liquid solution of glue had been used, which, when the texture of the MSS. was loose or broken, penetrated through three or four layers, and these, when the glue dried, separated together. To obviate this objection, I mixed the solution of glue with a sufficient quantity of alcohol to gelatinize it: and a mixture of the jelly and the fluid having been made and applied by a camel's hair brush, a film of jelly remained on the exterior of the surface of the leaf which attached itself to the membrane.

The effect of the solution of glue applied in the ancient method, was likewise to separate the layers, by expanding the imperfectly carbonized fibres. In the improvements I have mentioned, the alcohol, from its greater lightness, penetrated farther into the papyrus, but produced its greatest effect immediately on the first layers.

I adopted in some cases ether, as an agent for assisting the separation of the layers; and it was always found very efficacious, whether it was necessary to remove a single layer, or several layers at a time, in order to discover if a roll contained characters. The ether was applied by a camel's hair brush lightly to the surface of the leaf, when its operation was intended to be merely on that leaf; and it was suffered to sink deeper according as more layers were to be separated; the mere circumstance of its evaporation, which in some cases I assisted by heat, tended to detach the layers. For the black MSS. I employed sulphuric ether, and for the brown ones muriatic or nitric ether in their impure states, *i.e.* mixed with much alcohol.

No artificial modes had been employed by the Neapolitans for drying the papyrus in the operation of attaching the membrane, and no means, except mechanical ones, of detaching it after it was dried.

By throwing a stream of air, gradually warmed till it attained a temperature about that of boiling water, upon the surface of the leaf, I succeeded not only in drying the layers with much greater rapidity, but likewise in separating them with more delicacy.

I tried different modes of heating the air to be thrown upon the papyrus, such as passing it in a spiral metallic tube through warm water or oil by a double bellows, and from a large bladder through a straight tube having a very fine orifice and heated by a copper ball, surrounding the body of the tube, and exposed to burning charcoal; which last method, by its simplicity, I found the one best fitted to the Neapolitan operators. By sending the stream of air from a greater or smaller distance, so that it mixed with more or less cold air, the degree of temperature applied was regulated at pleasure. It was always found necessary to suffer a few minutes to elapse after the membrane was attached, and then to begin with a very slight increase of temperature; as otherwise, by too sudden an application of heat the membrane shrivelled before it became adherent, and the vapour suddenly raised destroyed its union with the papyrus; whereas, when the moisture was suffered to drain from the gelatinized glue, and the temperature was gradually raised, the expansion of the skin and the upper layer separated them perfectly from the lower layers, so that the unrolling was performed, as it were, by chemical means; and an operation, which hitherto had required some hours for its completion, was easily effected in from thirty to forty minutes.

I tried several experiments, by substituting solutions of resins in alcohol, and of gums in water, for the gelatinized solution; but none of them answered so well,—the resins would not adhere with any tenacity to the membrane; and the gums, when dried, had not that flexibility, which is an important character in the glue.

The alterations in the mode of applying and drying the membrane used to detach and preserve the leaves of MSS., capable of being unrolled, applied generally; I shall now mention the plans I adopted for the preparation of the MSS. for this operation.

MSS. in different states require a treatment of a directly opposite kind, which was to be modified according to circumstances. The pale chestnut-coloured MSS. covered partially with white ashes, were generally of a texture so loose, and had their layers so destroyed, that there was considerable danger of their falling into pieces by mere touching. The characters that remained in many of them were extremely distinct; and when a number of layers were taken up at once, it appeared as if they presented perfect columns of writing; but the fact is, the papyrus was full of holes, and each line was made up of letters from several different folds of the MS. When the process of unrolling these papyri was performed in the common way, the result obtained appeared, till it was examined minutely, a perfect column; but in fact it was made up of the letters of different words. I endeavoured to obtain the fragments of a single leaf attached to a layer of membrane, by applying a solution of caoutchouc in ether, to the surface of a MS., so as to supply the parts of the leaf destroyed; but, operating in this way, I obtained only a few characters, and never an entire word; so that, after various unsuccessful trials, I was obliged to give up the MSS. of this description as hopeless; more than five-sixths of their contents probably being always destroyed,—and that in so irregular a way, as to leave no entire sentences, or even words.

As to brown MSS., which were firm in their texture, and had the appearance of peat, and the leaves of which would not separate by common means, I tried the experiment of heating, after they had absorbed a small quantity of chlorine; and I found that in both cases the leaves detached themselves from each other, and were easily unrolled; but these MSS. had been so penetrated by water, that there were only a few folds which contained words, and the letters were generally erased, and the charcoal which had composed them, was deposited in the folds of the MSS.

Of the black MSS., of which the layers were perfect and easily separated, all the best specimens had been unrolled or operated upon, so that fragments only of this description remained. By assisting the operation of detaching the layers by muriatic ether, and the other processes mentioned in page 169, many parts of columns were obtained from several of the fragments, by which some idea of their contents may be formed.

On the black, compact, and heavy MSS. which contained white earthy matter in their folds, I tried several experiments, with the hopes of separating them into single layers, both by the action of muriatic and nitric ether, and by the operation of chlorine, and of weak hydrofluoric acid, assisted by heat; but generally the fibres of the papyrus had been so firmly cemented together, and so much earthy matter had penetrated them, that only a very imperfect separation could be obtained, and in parts where vestiges only of letters appeared,— so that from MSS. of this kind only a few remains of sentences could be gained.

During the two months that I was actively employed in experiments on the papyri at Naples, I had succeeded with the assistance of six of the persons attached to the Museum, and whom I had engaged for the purpose, in partially unrolling 23 MSS., from which fragments of writing were obtained, and in examining about 120 others, which afforded no hopes of success; and I should gladly have gone on with the undertaking, from the mere prospect of a possibility of discovering some better results, had not the labour, in itself difficult and unpleasant, been made more so, by the conduct of the persons at the head of this department in the Museum. At first every disposition was shown to promote my researches; for the papyri remaining unrolled, were considered by them as incapable of affording anything legible by the former method, or, to use their own word, *disperati;* and the efficacy and use of the new processes were fully allowed by the Svolgatori or unrollers of the Museum; and I was for some time permitted to choose and operate upon the specimens at my own pleasure. When, however, the Rev. Peter Elmsley, whose zeal for the promotion of ancient literature brought him to Naples, for the purpose of assisting in the undertaking, began to examine the fragments unrolled, a jealousy, with regard to his assistance, was immediately manifested; and obstacles, which the kind interference of Sir William A'Court was not always capable of removing, were soon opposed to the progress of our inquiries; and these obstacles were so multiplied, and made so vexatious towards the end of February, that we conceived it would be both a waste of the public money, and a compromise of our own characters, to proceed.

4. *Some general Observations.*

The Roman MSS. found in the Museum, are, in general, composed of papyrus of a much thicker texture than the Greek ones, and the Roman characters are usually larger, and the rolls much more voluminous; the characters of the Greek MSS., likewise, with a few exceptions, are more perfect than those of the Latin ones.

From the mixture of Greek characters in several fragments of Latin MSS., and from the form of the letters, and the state of decomposition in which they are found, it is extremely probable that they were of a very ancient date, when buried.

I looked in vain amongst the MSS., and on the animal charcoal surrounding them, for vestiges of letters in oxide of iron; and it would seem, from these circumstances, as well as from the omission of any mention of such a substance by Pliny, that the Romans, up to his period, never used the *ink of galls and iron* for writing: and it is very probable, that the adoption of this ink, and the use of parchment, took place at the same time.

For the ink composed of charcoal and solution of glue can scarcely be made to adhere to skin; whereas the free acid of the chemical ink partly dissolves the gelatine of the MSS., and the whole substance adheres as a mordant; and in some old parchments, the ink of which must have contained much free acid, the letters have, as it were, eaten through the skin, the effect being always most violent on the side of the parchment containing no animal oil.

The earliest MSS. probably in existence on parchment, are those codices rescripti, discovered by Monsignore Mai, in the libraries of Milan and Rome. Through his politeness, I have examined these MSS., particularly that containing some of the books of Cicero de Republica, and which he refers to the second or third century. From the form of the columns, it is very probable that they were copied from a papyrus. The vegetable matter, which rendered the oxide of iron black, is entirely destroyed, but the peroxide of iron remains; and where it is not covered by the modern MSS., the form of the letter is sufficiently distinct. Monsignore Mai uses solution of galls for reviving the blackness. I have tried several substances for restoring colour to the letters in ancient MSS. The triple prussiate of potash, used in the manner recommended by the late Sir Charles Blagden, with the alternation of acid, I have found successful; but by making a weak solution of it with a small quantity of muriatic acid, and by applying them to the letters in their state of mixture, with a camel's hair pencil, the results are still better.

It is remarkable, that no fragments of Greek, and very few only of Latin poetry, have been found in the whole collection of the MSS. of Herculaneum; and the sentences in the specimens we unrolled, in which Mr.

Elmsley was able to find a sufficient number of words to infer their meaning,* show that the works of which they are the remains were of the same kind as those before examined, and belonged to the schools of the Greek Epicurean philosophers and sophists.

Nearly 1000 columns of different works, a great part unrolled, under the superintendence of Mr. Hayter, and at the expense of his present Majesty George IV. have been copied and engraved by the artists employed in the Museum; but from the characters of the persons charged with their publication, there is very little probability of their being, for many years, offered to the world, which is much to be regretted; for though not interesting, from their perfection as literary works, they would unquestionably throw much light upon the state of civilization, letters, and science, of the age and country to which they belonged.

Should discoveries of MSS., at any future time, be made at Herculaneum, it is to be hoped that the papyri will be immediately excluded from the atmosphere, by being put into air-tight cases, filled with carbonic acid after their introduction, There can be no doubt but the specimens now in the Museum were in a much better state when they were first discovered; and the most perfect even, and those the coarsest in their texture, must have been greatly injured during the sixty-nine years that they have been exposed to the atmosphere. I found that a fragment of a brown MSS., kept for a few weeks in a portion of air confined by mercury, had caused the disappearance of a considerable part of the oxygen, and the formation of much carbonic acid.

* Engravings of copies of a few of these fragments, selected from nearly 100, are annexed to this paper, for the purpose of shewing their nature.

PLATE 3.

Fig. 1.

Fig. 2.

PLATE 4.

Fig. 1

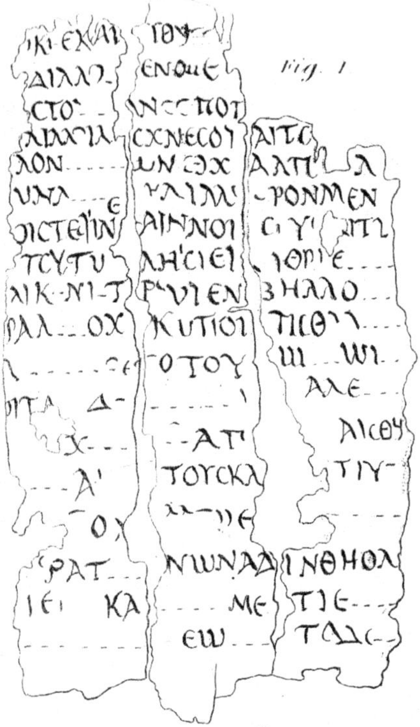

Fig. 2

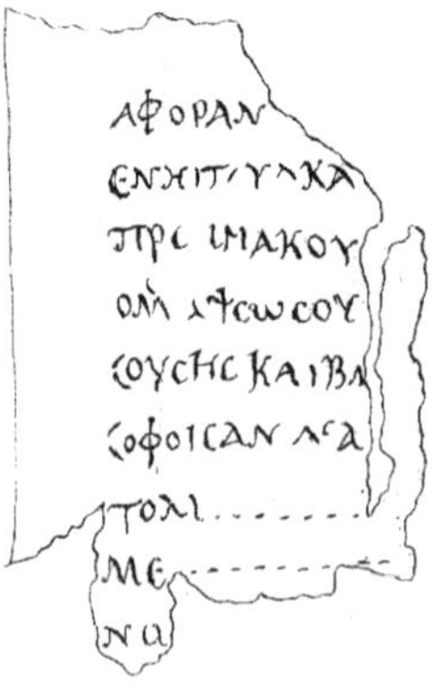

Fig. 3

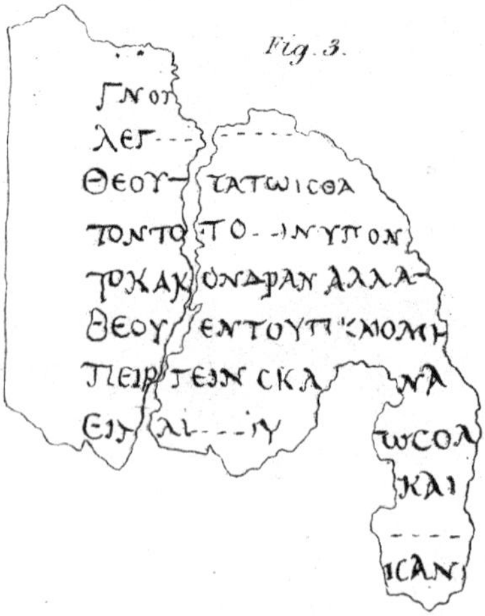

PLATE 5.

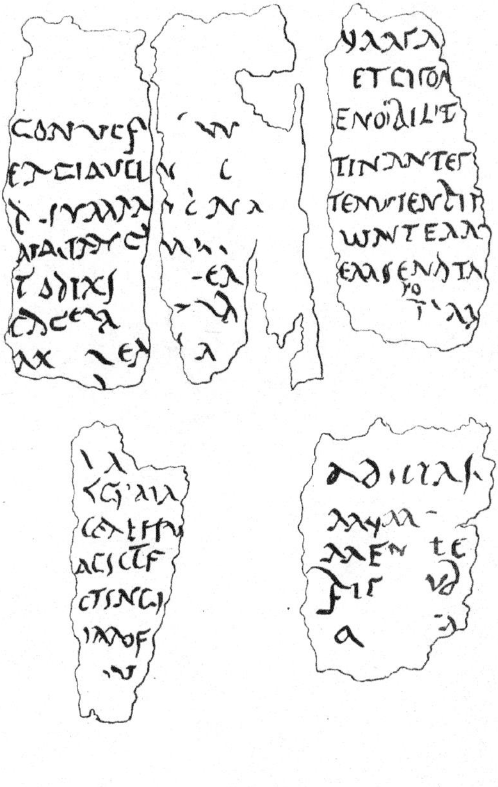

PLATE 6.

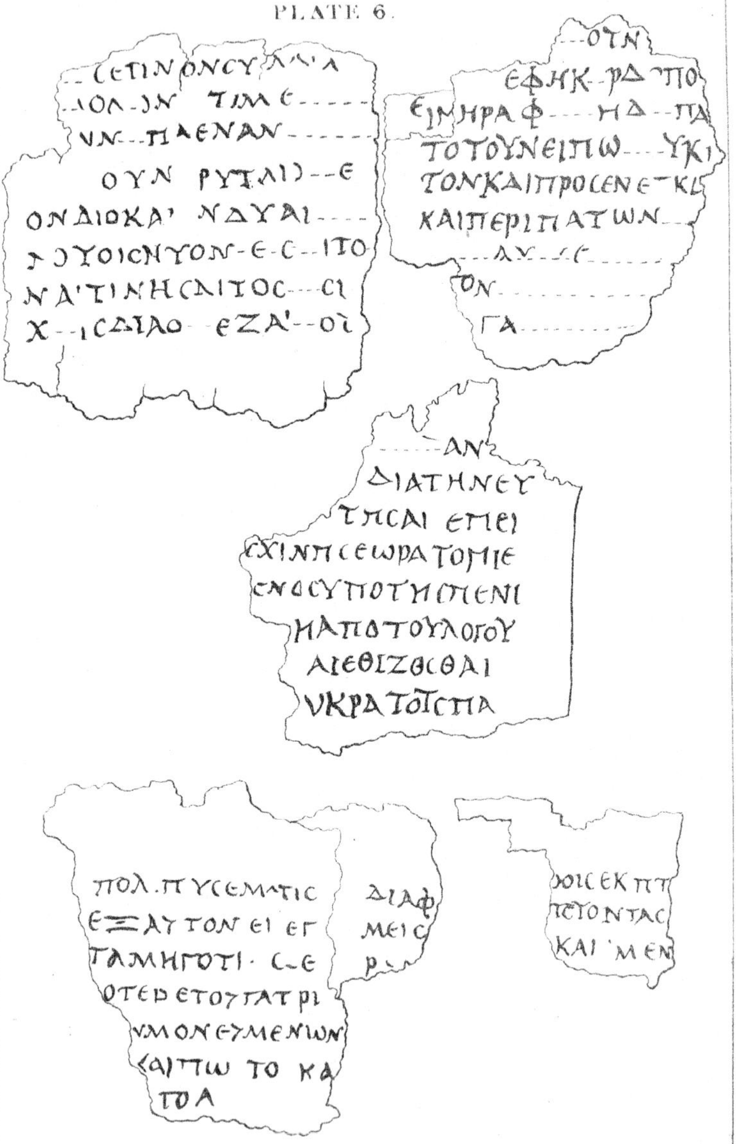
ΤΟΤΟΥΝΕΙΠΩ
ΤΟΝΚΑΙΠΡΟCΕΝΕ
ΚΑΙΠΕΡΙΠΑΤΩΝ
ΔΙΑΤΗΝΕΥ
ΗΑΠΟΤΟΥΛΟΓΟΥ
ΑΙΕΘΙΖΕCΘΑΙ
ΥΚΡΑΤΟΥCΠΑ
ΥΜΟΝΕΥΜΕΝΩΝ
ΤΕΤΟΝΤΑC
ΚΑΙ ΜΕΝ

PLATE 7.

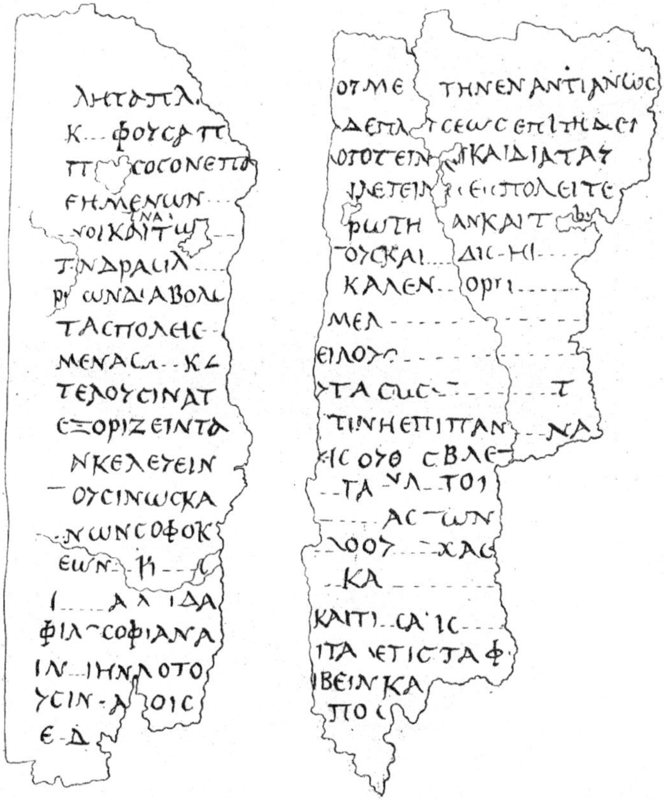

PLATE 8.

CONSI
CLEB
ORDVS
INESSE
QVID
ORDI
NO
TV
MO
IAC
VMQ
IVR
IROIECII

PLATE 9.

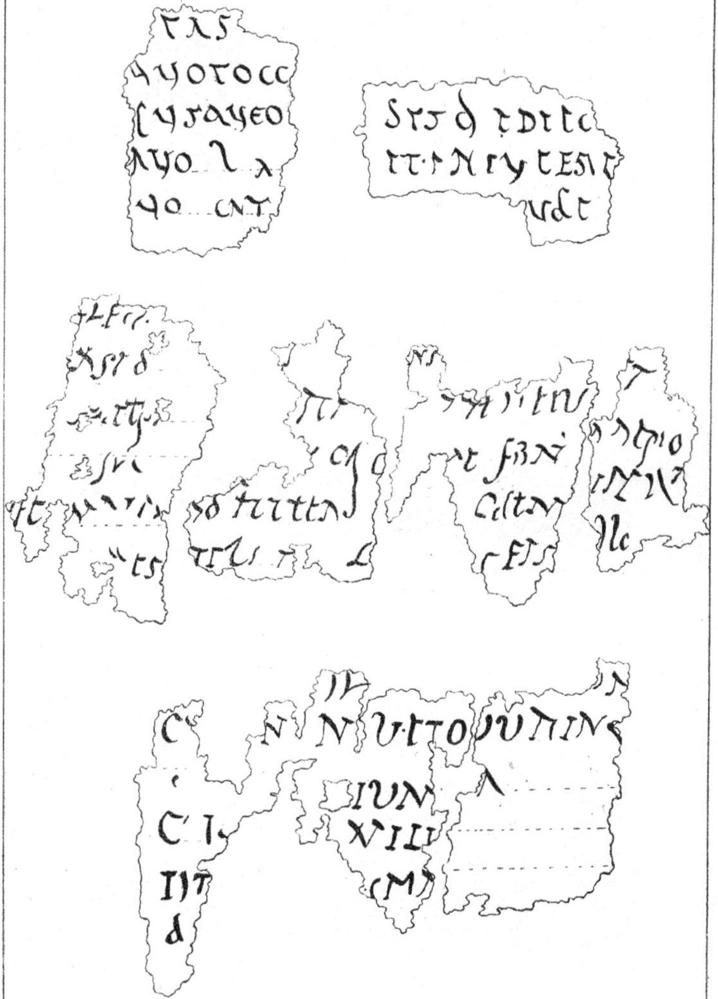

PLATE 10.

Fig. 1.

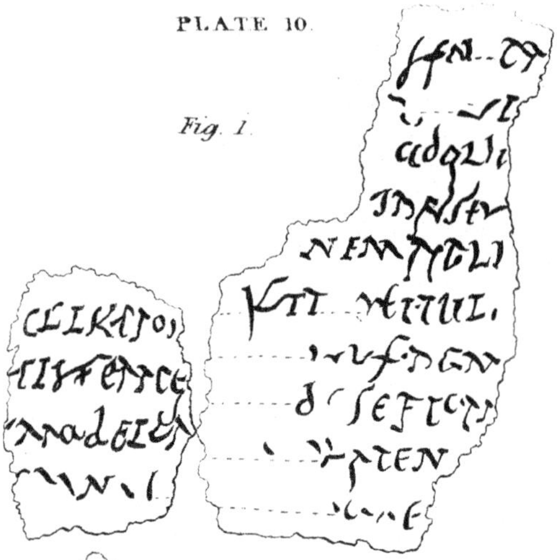

Fig. 2.

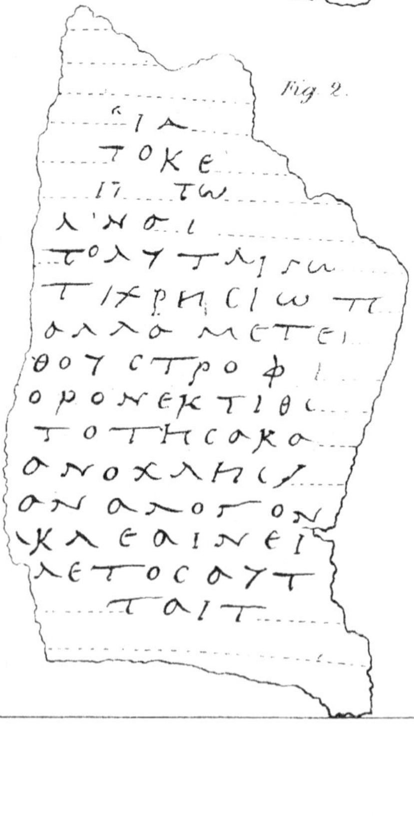

Plate III. Fig. 1. represents a papyrus, partly unrolled, with the ink-stand and reed for writing, used by the ancients.

Fig. 2. represents a box of papyri; both copied from the " Pitture antiche d'Ercolano."

Plate IV. fig. 1. is a specimen of an unrolled papyrus, which is so destroyed, that the letters of different columns appear through the folds, as if they formed one column.

Figs. 2 and 3. are specimens of fragments, in which the lines begin with Greek capitals.

Plate V. contains a specimen of a fragment of a Roman MS., of which the characters are partly Greek.

Plate VI. contains specimens of fragments of another Greek MS.

Plate VII. contains specimens of fragments of another Greek MS.

Plate VIII. contains specimens of fragments of a MS. in Roman capitals.

Plate IX. contains specimens of MSS., supposed to be Roman, written in peculiar characters.

Plate X. fig. 1. specimen of a fragment of a MS. of which the characters have not yet been examined.

Fig. 2. specimen of a fragment of Greek MS.

IV.

ON THE CAUSE OF THE DIMINUTION OF THE TEMPERATURE OF THE SEA ON APPROACHING LAND, OR IN PASSING OVER BANKS IN THE OCEAN.*

In the third number of the Journal of Science and the Arts, an extract of a letter from my brother, Dr. Davy, has been published, containing some observations on the temperature of the ocean and the atmosphere, in the equatorial regions of the globe. I have since communicated to the Royal Society a long letter which he has written on this subject, and which will be published in the next volume† of the Transactions. Amongst other philosophical remarks, those by which he confirms the conclusion of Mr. Jonathan Williams,‡ and other observers, that the temperature of the sea always falls in shoal water, and that the thermometer may be made a useful instrument in navigation, appear to be very important.

Mr. Williams attributes the effect upon the thermometer, on the approach of land, to the cooling power of the land; but this reason will not apply to the effect of shoals in the ocean, or to the tropical climates. M. de Humboldt, in his Personal Narrative, seems to consider it as resulting from cold currents below the surface; but in his work he does not enter into any minute

* [From the Journal of Science and the Arts, vol. iii. 1817.]

† [Phil. Trans. for 1817.]

‡ Thermometrical Navigation, Phil. 1779.

details; nor in a conversation which I had with him on the subject, did he pursue it any further, than in mentioning this general opinion. Dr. Davy has merely noticed the fact, as a general law; but has not speculated upon the cause of it.

The great interests of the subject to the practical navigator, has induced me to consider the theory of it rather minutely, and I shall now detail my views. The same reasons, I have no doubt, either have occurred, or will occur to M. de Humboldt and to Dr. Davy; but I am sure no apology will be necessary for anticipating these, either to my brother, or to the illustrious Prussian traveller, whose candour and urbanity are equally distinguished with his knowledge and sagacity.

The solar rays produce very little heat, in passing through the air; but during their transmission through a body so imperfectly transparent as water, there can be no doubt that the same cause which occasions a loss of light, must communicate an effect of heat, and consequently the greatest heat must be produced at the surface of the sea,* and it must gradually diminish, as the rays penetrate deeper.

* [This is strongly indicated during a calm; in the paper of mine, referred to in the Philosophical Transactions for 1817, an instance in point is given; although the calm was not of twenty-four hours' duration, the temperature of the sea at the surface rose more rapidly than that of the air; an extract may be given in proof.

"August 7th, Lat. 2′ 10′ N. Long. 76° 47′ E.

Hour.	Temp. of air.	Temp. of water.
6 A. M.	78·5	80°
8 —	79·5	81
10 —	80·5	81·5
12 —	82·	82·5
2 P. M.	82·5	83·5
3 —	82·	83·5

Further, it may be mentioned in corroboration, that according to the

The heat of the surface of the ocean must, at a great distance from land, depend upon the absorption of the solar rays, the cooling of this surface upon its radiating powers, and upon evaporation. But water is an imperfect conductor of heat; and by cooling, as far as 38° or 40° of Fahrenheit, its density is increased: when cooling agencies act upon an unfathomable ocean, the strata of water *cooled* sink out of the reach of the surface, and very little influence the temperature of this surface; but when cooling agencies act upon a shallow part of the ocean, the cooled strata accumulate and approach nearer the surface, and cause the temperature of the ocean, at its surface even, to be nearer the mean temperature of day and night.

In *very shallow* water, close to the shore, the bottom will be heated; and during the day the temperature, close to the shore, will be higher than that of the ocean; but in the night, as the land* cools faster than the sea by radiation, the air, having its temperature lowered by contact with the cooled land, will flow down upon the sea, and thus will destroy the effect of the hot water flowing from the extreme shallows,† and at a certain

most accurate thermometrical observations which have yet been made, the temperature of the ocean within the tropics diminishes in a certain ratio with the depth, until, at a certain depth, it is of that degree at which the fluid is of its maximum density—namely, about 40° of Fahrenheit.]

* [This remark, I believe, is applicable to the surface of a shoal; it radiates more heat than the water incumbent on it; and consequently exercises a special cooling agency,—diminishing the temperature of the water in contact with it.]

† [Where shallows are extreme, they rapidly become hot under the influence of sunshine, and rapidly cool after sunset: this may be illustrated by the experiments of exposing a bottle of water to the open sky;—at Malta I have seen a wine bottle so filled, placed on the terrace-roof of a house on cotton-wool, rise by day to 110°, and fall by night to 69·5, at a time when the day temperature of the air was about 84°, and the night temperature about 73°; the one the maximum, the other the minimum.]

moderate distance will produce such a diminution of temperature, as will more than compensate for the heat produced by contact with hot land. Hot-air and water, within limits above 52°, always rise; cool air and water above 40°, sink; therefore, by whatever cause cool air or water are kept near the surface of the ocean, that cause will diminish the general temperature of that surface.

It has been supposed by M. Perron, and other inquirers, that ice may exist at the bottom of the ocean; but simple physical reasons show that this is impossible, unless the temperature of the surface of the ocean is below 40°; for water at 40°, is heavier than at the freezing point. Ice, as Count Rumford has shown, always forms at the surface: * and ice at the bottom of any part of the ocean, must begin to thaw, when the temperature of the surface is above 40°: for hot currents then descend, and cold ones rise.

The same cause must always operate, when the heat of the surface of the ocean is above 52°; and under these circumstances, whether in the equatorial, polar, or tropical climates, land or shallows must always lower the temperature of the ocean: but in very high latitudes, if the heat of the surface should approach to 40° only, the thermometer will no longer be a guide of land to the navigator, for water is heavier at 47°, than at the freezing point 32°; but such an occurrence can only happen in icy seas.

* [Occasionally ice forms below the surface, at the bottom; the occurrence of ground-ice appears to be connected with the cooling effects of radiation from the bottom, and the mass of flowing water having been reduced to 32°, or to nearly that degree.]

V.

SOME OBSERVATIONS ON THE FORMATION OF MISTS IN PARTICULAR SITUATIONS.*

All persons who have been accustomed to the observation of nature, must have frequently witnessed the formation of mists over the beds of rivers and lakes in calm and clear weather after sun-set; and whoever has considered these phenomena in relation to the radiation and communication of heat and the nature of vapour, since the publication of the researches of MM. Rumford, Leslie, Dalton, and Wells, can hardly have failed to discover the true cause of them. As, however, I am not aware that any work has yet been published in which this cause is fully discussed, and as it involves rather complicated principles, I shall make no apology for offering a few remarks on the subject to the Royal Society.

As soon as the sun has disappeared from any part of the globe, the surface begins to lose heat by radiation, and in greater proportions as the sky is clearer; but the land and water are cooled by this operation in a very different manner: the impression of cooling on the land is limited to the surface, and very slowly transmitted to the interior; whereas in water above 45° Fahrenheit, as soon as the upper stratum is cooled, whether by radiation or evaporation, it sinks in the

* [From the Phil. Trans. for 1819. Read before the Royal Society, Feb. 25, 1819.]

mass of fluid, and its place is supplied by warmer water from below, and till the temperature of the whole mass is reduced to nearly 40° F. the surface cannot be the coolest part. It follows, therefore, that wherever water exists in considerable masses, and has a temperature nearly equal to that of the land, or only a few degrees below it, and above 45° F. at sun-set, its surface during the night, in calm and clear weather, will be warmer than that of the contiguous land; and the air above the land will necessarily be colder than that above the water; and when they both contain their due proportion of aqueous vapour, and the situation of the ground is such as to permit the cold air from the land to mix with the warmer air above the water, mist or fog will be the result; which will be so much greater in quantity, as the land surrounding or inclosing the water is higher, the water deeper, and the temperature of the water, which will coincide with the quantity or strength of vapour in the air above it, greater.

I shall detail some observations which appear to me to show the correctness of this view. June 9th, 10th, 11th, the temperature of the atmosphere and of the Danube was repeatedly examined during a voyage that I made upon this river from Ratisbone to Vienna, and on each of these days, the sky being perfectly clear, the appearance of mist above the river in the evening uniformly coincided with the diminution of the temperature of the air from three to six degrees *below* that of the river, and the disappearance of fog in the morning with the elevation of the temperature of the air *above* that of the river. From Ratisbone to Passau, the temperature of the Danube was pretty uniform throughout the 24 hours, being highest, 62° F. or 62¼° F., between 12 and 2 o'clock, and about one degree less before sun-rise, and the temperature of the air from 61° F. to 73° F. during the day, and from 61° to 54° F. during the night. Below Passau, the Inn and Ilz flow into the Danube.* On examining the temperature of these rivers at 6 o'clock, A.M., June 11, that of the Danube was found to be 62° F., that of the Inn 56½° F., and that of the Ilz 56° F.: the temperature of the atmosphere on the banks where their streams mixed, was 54°. The whole surface of the Danube was covered with a thick fog; on the Inn there was a slight mist, and on the Ilz barely a haziness, indicating the deposition of a very small quantity of water. About 100 yards below the place where the three rivers joined, the temperature of the central part of the Danube was 59° F., and here the quantity of mist was less than on the bed of the Danube before the junction; but about half a mile below, the warmer water had again found its place at the surface, and the mist was as copious as before the union of the three rivers. June 12th, the evening was cloudy, and the temperature of the atmosphere remained till after dark higher than that of the river, being, when the last observation was made, 63° F., when there was not the slightest appearance of mist. The sky was clearer before sun-rise on the 13th, and the thermometer immediately after sun-rise, in the air above the river, stood at 55½° F., the temperature of the Danube being 61° F.; a thin mist was seen immediately above the river; but there being no mass of vapour to exclude the sun-beams, it rapidly disappeared; and was not visible a few feet from the surface; and in half

* The Danube was greenish, the Inn had a milky blueness, the Ilz was perfectly pellucid; but from the rapidity with which the Inn descended, its waters at this spot give their tint to the whole surface.

an hour the whole atmosphere was perfectly transparent.

In passing along the Rhine from Cologne to Coblentz, May 31st, and June 2nd and 3rd, the nights being very clear, the same phenomenon of the formation of mists was observed, precisely under the same circumstances; but as I could examine the temperature of the air and the river only close to the banks, and in two or three situations, my observations were less precise and less numerous; the mist formed later in the evening, and disappeared sooner in the morning than on the Danube; which was evidently owing to the circumstances of the atmosphere being warmer, and the river colder, the temperature of the one being from 66° F. to 75° F., during the day, and that of the river, where I examined it, from 59° to 60° F.

July 11th I examined the temperature of the Raab, near Kermond, in Hungary, at 7 o'clock, P.M., and found it 65° F., that of the atmosphere being 72° F. During the whole evening there were some thin fleecy clouds in the western sky, which being lighted up by the setting sun, greatly interfered with the cooling by radiation from the earth, and at half-past nine, the thermometer, in the atmosphere, was still 69° F., and at half-past ten 67° F., when there was not the slightest appearance of mist. In the morning, before sun-rise, the temperature of the atmosphere on the banks was 61° F., that of the river 64° F., and now the bed of the river was filled with a white thin mist, which entirely disappeared half an hour after sun-rise.

I made similar observations on the Save, in Carniola, in the end of August; on the Izonzo in the Friul, in the middle of September; on the Po near Ferrara, in the end of September; and repeatedly on the Tiber, and on the small lakes in the Campagna of Rome, in the beginning of October; and I have never in any instance observed the formation of mist on a river or lake, when the temperature of the water has been lower than that of the atmosphere, even when the atmosphere was saturated with vapour.

It might at first view be supposed, that whether the cooling cause existed in the water or the land, the same consequences ought to result; but the peculiar properties of water, to which I referred in the beginning of the paper, render this impossible. Water in abstracting heat from the atmosphere becomes lighter, and the warmer stratum rests on the surface, and its operation in cooling the atmosphere is extremely slow; besides, the cooled atmospheric stratum remains in contact with it, and water cannot be deposited from vapour, when that vapour is rising into an atmosphere of a higher temperature than its own; and the law holds good however great the difference of temperature. Thus, August 26th, at sun-set, the day after a heavy fall of rain, and when the atmosphere was exceedingly moist, I ascertained the temperature of the Drave, near Spital, in Carinthia, and though it was 14° F. below that of the air, yet the atmosphere above the river was perfectly transparent.

It may be imagined, but without any reference to the cooling agencies of air from the land, mist may form upon rivers and lakes, merely from the loss of heat by radiation from the air, or the vapour itself immediately above the water; and that the phenomena is merely one of the formation of vapour, the source of heat being in the water, and its deposition, the source of cold, being in the atmosphere; but it is extremely improbable that air or invisible vapour, at common temperatures, can lose any considerable quantity of heat by radiation; and if mist could be formed from such a source, it must always be produced to a great extent upon the ocean in calm weather during the night, particularly under the line and between the tropics, which the journals of voyages sufficiently prove is not the case. I have myself had an opportunity of making some observations which coincide with this view. During a voyage to and from Pola, I passed the nights of September 3, 5, and 6, off the coast of Istria; there was very little wind on either of the nights, and from sun-set till nearly midnight it was perfectly calm in all of them. On the 3rd it was cloudy, and the lightning was perceived from a distant thunder-storm, and the vessel was never far from the shore; but on the 5th and 6th the sky was perfeetly clear, and the zodiacal light, after sunset, wonderfully distinct and brilliant, particularly on the 5th, and we passed, by the help of oars, from two to eight miles from the shore. The temperature of the sea at sun-set was 76° F. on the 5th, 77° F. on the 6th, that of the atmosphere immediately above it 78° F. and 79° F. On the 5th at midnight, about five miles from the shore, the temperature of the sea was 74° F., and that of the atmosphere 75° F. There was not the slightest appearance of mist on either of these nights on the open sea, or at any distance from the land; but close under the hills of Istria there was a slight line of haze visible before sun-rise, which was thickest under the highest land; and as we approached, at sun-rise on the 7th, the mountains of the Friul, the tops of those nearest to Trieste were seen rising out of a thick white mist, which had not reached a quarter of a mile from the shore.

After mists have formed above rivers and lakes, their increase seems not only to depend upon the constant operation of the cause which originally produced them, but likewise upon the radiation of heat from the superficial particles of water composing the mist, which produces a descending current of cold air in the very body of the mist, whilst the warm air continually sends up vapours: it is to these circumstances that the phenomena must be ascribed of mists from a river or lake, sometimes arising considerably above the surrounding hills. I have often witnessed this appearance during the month of October, after very still and very clear nights, in the Campagna of Rome above the Tiber, and on Monte Albano over the lakes existing in the ancient craters of this extinguished volcano, and, in one instance, on the 17th of October, before sun-rise, there not being a breath of wind, a dense white cloud of a pyramidal form was seen on the site of Alban lake, and rising far above the highest peak of the mountain, its form gradually changed after sunrise, its apex first disappeared, and its body, as it were, melted away in the sunbeams.

Where rivers rise from great sources in the interior of rocks or strata, as they have the mean temperature of the climate, mists can rarely form upon them, except in winter, or late in autumn, or early in spring. In passing across the Apennines, October 1st, 2nd, and 3d, 1818, there having been much rain for some days preceding, and the nights being very clear, I observed the beds of all the rivers in the valleys filled with mist, morning and evening, except that of the Clitumnus near its source, in which there was no mist, and this river rises at once from a limestone bed, and when I examined it, at half-past six o'clock, A.M., October 3, was $7\frac{1}{2}$ lower than the atmosphere.

Great dryness in the air, or a current of dry air passing across a river, will prevent the formation of mist, even when the temperature of the water is much higher than that of the atmosphere; thus, on the 14th of June, near Mautern, though the Danube at five in the morning was 61° F. and the air only 54°, yet there was no mist; but a strong easterly wind blew, and from the rapidity with which water evaporated, it was evident that this wind was in a state of extreme dryness.

The Tiber has furnished me with a number of still more striking examples. October 13th, the night having been very clear, on arriving at the Ponte Molle, at half-past six in the morning, I found no mist on the river, yet the temperature of the air immediately above it was 48° F., and that of the river 56° F.; a strong north wind blew, which indicated, by the hygrometer, a degree of dryness of 56°, and this part of the river was exposed to it; but the valley above, where the river was sheltered from the wind, was full of mist, and the mist in rising to the exposed level might be seen, as it were, dissolving, presenting thin striæ which never reached above a certain elevation, and many of which disappeared a few seconds after they rose. From the 13th to the 25th of October, during which time the tramontane or north wind blew, I witnessed repeatedly the same phenomena, and in the whole of this time there was only one morning when there was no mist in the sheltered valley, and the cause was perfectly obvious; the night had been very cloudy, and the thermometer, before sun-rise, indicated a difference of only one degree in the atmosphere below that of the river.

It is not my intention to discuss the general subject of the deposition of water from the atmosphere in this paper; but merely to describe a local cause of considerable extent and variety in its modifications, and which is not without an effect in the economy of nature;

for verdure and fertility, in hot climates, generally follow the courses of rivers, and by the operation of this cause, are extended to the hills, and even to the plains surrounding their banks.

Rome, Dec. 8th, 1818.

VI.

OBSERVATIONS ON THE NATIVE CAUSTIC LIME OF TUSCANY.*

The Duchess of Montrose was so good as to send me the caustic lime, which is the subject of the analysis [referred to below.] Her Grace received it immediately from Tuscany. It was in a bottle, carefully sealed and full of water. Some of the exterior portions had become combined with carbonic acid before they were collected, and from the colour, it appeared that there were different portions of protoxide of iron in different parts of the substance.

On examining the water it was found to be a saturated solution of lime and it contained fixed alkali, but in quantities so minute, that after the lime was separated, it could be made evident only by coloured tests.

It appears from Mr. Faraday's analysis, that the

* [From the Journal of Science and the Arts, edited at the Royal Institution, vol. i. 1816. This paper was preceded by one by Mr. Faraday, (I believe his first contribution to science,) giving the following results of his analysis of the deposit in question :

Lime	82·424
Silex	10.57
Iron	2·82
Alumine	1·34
Loss	2·846
	100.000

From a statement of the Marquis Ridolfi in the same Journal, it would appear that the caustic lime of Tuscany is found between a superficial stratum of carbonate of lime, and the clayey bottom of a pool (laguna,) the ancient bath of Santa Gonda, and is derived from two springs, abounding in the earth, one of which is hot.]

menstruum which deposits the solid substance must be a solution of silica in lime-water; and heat is evidently the agent by which the large quantity of lime deposited is made soluble, and is enabled to act on silica; and the fact offers a new point of analogy between the alkalies and the alkaline earths.

Vestiges of extinct volcanoes exist in all the low countries on the western side of the Apennines, and the number of warm springs in the Tuscan, Roman, and Neapolitan states, prove that a source of subterraneous heat is still active in a great part of the surface in these districts.

Carbonic acid gas is disengaged in considerable quantities in several of the springs at the foot of the Apennines; and some of the waters that deposit calcareous matter are saturated solutions of this substance. Calcareous tufas of recent formation are to be found in every part of Italy. The well known Travertine marble, Marmor Tiburtinum, is a production of this kind; and the Lago di Solfaterra, near Tivoli, of which I shall give a particular account on a future occasion,* annually deposits masses of this stone of several inches in thickness.

It is scarcely possible to avoid the conclusion that the carbonic acid, which by its geological agency has so modified the surface of Italy, is disengaged, in consequence of the action of volcanic fires on the limestone, of which the Apennines are principally composed, and liberated at their feet, where the pressure is comparatively small; but the Tuscan laguna offers the only instance in which the action of these fires extends, or has extended to the surface at which the water collected in the mountains finds its way to the sea, so as to enable it to dissolve caustic calcareous matter.

* [Consolations in Travel, Dialogue III.]

VII.

HINTS ON THE GEOLOGY OF CORNWALL.*

I SHALL in this letter comply with the request that you did me the honour to make, of offering a few hints respecting the geology of Cornwall. I can hardly venture to hope that they will be worth the attention of the Society; most of the members have had much better opportunities than have occurred to me, of examining our interesting county; and I dare say that many of the observations I shall make, will have been anticipated by others: at all events, this communication will show the desire I have to co-operate in promoting the useful objects of the Society, and I trust they will consider it as a proof of my respect.

Cornwall may be regarded κατεξοχην as the *country of veins.* It is in veins that the most useful, as well as the most valuable minerals, generally exist; that the pure specimens are found, which serve to determine the mineralogical species, and that the appearances seem most interesting in their connexion with geological theory. Thus veins, which may be considered in the light of the most valuable cabinets of nature, were once her most active laboratories; and they are equally important to the practical miner, and to the mineralogical philosopher.

* [From the Transactions of the Royal Geological Society of Cornwall, vol. i. 1818. In a letter to H. Borse, Esq. Treasurer.]

Amongst the veins of Cornwall most curious, in a geological point of view, are those of *granite.* These formations are extremely numerous. As far as I am acquainted with them, they either intersect micaceous schist, or other granite rocks.

The most remarkable instances known to me in Cornwall, are those on the south-east extremity of St. Michael's Mount, those near Mousehole Cove, those at Zennor, and on the east and west side of Cape Cornwall.* At St. Michael's Mount, the granite veins contain fragments of micaceous schist; and whoever examines without prejudice the different formations, cannot, I think, doubt that the vein has been produced in a chasm of the rock in which it occurs, and that it is, consequently, *posterior* in formation.

Similar instances of veins of granite, are described by mineralogists; a very interesting one was shown to me at Killiney, near Dublin, by Dr. Blake: some in the Isle of Arran, and in other parts of Scotland, are well known. I have seen several cases of granite veins near Morlaix, in Britanny. I do not know that any analogous formations have been observed in the great mountain chains of Europe. I have looked for them in vain in the points of junction of the schist with granite, both in the Maritime, Savoy, Swiss, and Tyrolese Alps, and likewise in the Oriental Pyrenees. My researches have not been extensive or minute; but I should be disposed to conclude from what I have seen, that granite veins are peculiar to the low metalliferous granite and mica schist formations.

* I first observed the granite veins of Cornwall about eighteen years ago; probably before, and certainly to a great extent since, they have occupied the attention of geologists; and they are too well known to the Geological Society, to require any topographical history from me.

There are deposited in the Museum of the Royal Institution, three specimens of fragments of micaceous schist, included in granite, which are from Cornwall; two from St. Michael's Mount, and one from St. Just. At the points of junction of micaceous schist with the low granite, in various cases which I have seen, rocks of granite occur, containing considerable portions of the schist; this is particularly the case at Killiney in Ireland, and near Balyhulish in Scotland. I have seen similar instances in the granite used in building in Mayence, in France; and in a pillar of granite, which was erected some years ago to Buonaparte at Marseilles, there is a very large fragment of mica schist.

The place where the granite joins the schist in St. Michael's Mount, is remarkable for the number of crystallized substances it contains; oxide of tin, wolfram, phosphate and fluate of lime, quartz, mica, felspar, topaz, are all well known to occur in the same veins; and the particular investigation and history of the nature of the crystallized bodies occurring in this beautiful and remarkable spot, appear well worthy of the attention of the Geological Society, established so immediately in its neighbourhood.

An opinion has been expressed by a foreign naturalist, who was extensively employed in geological researches, some time ago by the Geological Society of London, that the granite veins of Cornwall are mere protuberances of primary granite, in which mica schist was found. This opinion does not merit discussion, and could only have been formed in consequence of very superficial examination. It might, with nearly as much reason, be stated, that the veins of copper and tin belong to a great interior metallic mass, and that they existed prior to the rocks in which they are found. The notion that these veins are contemporaneous with the rock, is more plausible; and a forced explanation of many of the phenomena, might be given on this view; but it is contradicted by the fragments of mica schist found in the veins.

The porphyry of Cornwall in general, seems to belong to the dyke or vein formation; and its history is an object of considerable geological interest. The northern shores of Mount's Bay, exhibit great varieties of this substance; and it contains a number of crystallized substances, which intersect it in veins. A most remarkable vein of this kind, was worked some years ago at the Wherry mine, near Penzance; the principal metals were oxide of tin, and sulphuret of copper; but ores of cobalt and lead likewise occurred; and the variety of metallic substances found with them, in minute quantities, was very extraordinary. I have seen in the refuse heaps, blende, oxide of uranium, oxide of titanium, and of iron, pech-blende, nickel and arsenical pyrites; and, in a single piece of the vein, of a few inches square, many of these substances might be found embedded in quartz or chlorine.

A very good account of the working of this mine has been drawn up by Mr. Hawkins, one of the members of our Society, and published in German; and I have seen a French translation of it in the Journal des Mines. This paper, in an English dress, ought to be placed in the archives of the Geological Society of Cornwall; and is worthy of being inserted in their first publication, as a record of the industry and ingenuity with which great natural obstacles were overcome.* The Wherry mine is the only instance, I believe, that has ever oc-

* [A very interesting account of this mine is to be found in the same volume, written by Mr. Hawkins, under the title of Submarine Mines.]

curred of a shaft sunk in the ocean; and of a mine worked, the only access to which, is below low-water mark.

The *serpentine* district of Cornwall has not yet met with the attention it deserves. I have seen no formation in which the nature of serpentine is so distinctly displayed. The true constituent parts of this rock appear to be resplendent hornblende and felspar; it seems to differ from sienite only in the nature of the hornblende, and in the chemical composition of its parts, and in being intersected by numerous veins of steatite and calcareous spar. Near Coverack Cove, the felspar and resplendent hornblende forming the rock, occur in crystals of some inches in size; and from this size there is a gradation to crystals so minute, that the rock appears of a simple nature. The green or red colour of the hornblende is generally the cause of the peculiar tints of the rock; the felspar is generally white, but, in instances at Coverack, some of the large crystals of felspar are of a reddish hue.

The nature and origin of the veins of steatite in serpentine are curious subjects of inquiry. Were they originally crystallized, and the result of chemical deposition?—or have they been (as, for the most part, they are now found) mere mechanical deposits? I am inclined to the last opinion. The felspar in serpentine is very liable to decompose, probably from the action of carbonic acid and water on its alkaline, calcareous and magnesian elements; and its parts washed down by water, and deposited in the chasms of the rock, would necessarily gain that kind of loose aggregation belonging to steatite.

I made, some years ago, a rude comparative analysis of the felspar in serpentine, and of the soap rock. I found the same constituents in both of them, except that there was no alkali nor calcareous earth in the steatite; but my experiments were not so exact as to determine the proportions. It is not easy to conceive that steatite was originally a crystallized substance, which has been since decomposed; for in that case it ought to be found in its primitive state, in veins excluded from the action of water and air; and it is not difficult to account for the hardness of some species of steatite, on the hypothesis which I have stated. Mere mechanical deposits, when very finely divided, and very slowly made, adhere with a considerable degree of force. A remarkable instance of this kind has occurred to me: amongst the chemical preparations of the late Henry Cavendish, Esq. which were given me by Lord George Cavendish, there was a bottle which had originally contained a solution of silica by potassa; the cork had become decayed during the lapse of years, and carbonic acid of the atmosphere had gradually precipitated the earth, so that it was found in a state of solid cohesion; the upper part was as soft as the steatite of the soap-rock, but the lower part was very hard, was broken with some difficulty, and had an appearance similar to that of chalcedony.

The felspar of serpentine seems to differ from that of other rocks, in containing a much larger proportion of magnesian earth; but many varieties of felspar are liable to decomposition; the porcelain clay of St. Stephens is well known to be the result of a process of this kind.

I have seen a specimen of porcelain clay from a mine in the west of St. Just, which contained a quantity of magnesia, and which appeared to be produced by the disintegration of felspar. It occurred in a quartzose vein, which afforded oxide of tin. Copper, I believe, is the only metallic substance that has been found in any quantity in the serpentine formation of Cornwall, and crystallized substances in general are rare in the cavities of this rock. In America, chromate of iron is found in serpentine; and in the serpentine of the coast of Genoa, Professor Viviani has shewn me arragonite, and a peculiar crystallized stone analagous to crysolite. The relation of the Cornish serpentine to the neighbouring rocks is worthy of examination; towards the north it seems to pass into micaceous schist; at the point of Cape Lizard it is bounded by stratified sienite, and white micaceous schist. In Kynan's Cove, so remarkable for the beauty of the forms of its cliffs, and for the number of its caverns, there is a protuberance of granite above the white sand. I do not believe that the Cornish serpentine has, as yet, been applied to any purposes of architecture or sculpture, and, in general, at the surface, its parts are too small, and contain too many fissures to be worked with advantage; but I am convinced, that by making proper excavations, many parts of the serpentine district would afford large and beautiful blocks of great fineness and beauty of colour.

The serpentine of the Apennines is, in general, very like that of Cornwall. During an examination of the coast of Italy, that I made between Genoa and Massa, I found, a few miles from Sestri di Levanto, an ancient quarry, bearing marks of having been worked by the Romans; but which, amongst the profusion of marble furnished by the neighbourhood of Carrara, and amidst the quantities of ancient remains of verde antique, is neglected by the modern Italians.

I hope some members of the Society will examine the black marble of the south coast of Cornwall. From a superficial examination that I made of a part of this district, it appeared to me to be of the same formation as the Plymouth marble.

I am ignorant whether any true basalt exists in Cornwall. Near Port Isaac I have seen on the road wacké, or mandelstein, but I do not know where it was quarried. A sienite, resembling the Scotch grunstein, is very common on the eastern border of the county; but I am inclined to believe that it belongs to the primary formation.

Though the whole of Cornwall is of a very peculiar mineralogical construction, compared with the rest of Britain, and even of Ireland, yet it bears a very considerable relation to the opposite coast of France. The mica and chlorite schist, and the granite (the killas and growan) in the neighbourhood of Morlaix, in Brittany, I find precisely similar to those of Mount's Bay, and containing similar kinds of actinolite and thumerstein; and I saw at Morlaix specimens of serpentine (said to have been brought from the neighbourhood of Rosloff,) similar to those of Cape Lizard. Veins of tin have been worked in the low French granite formation; but as yet they have not been productive.

The conformation of Cornwall, is in the highest degree curious; and the facts it offers, are illustrative of many important points of geological theory. It exhibits very extraordinary instances of rocks broken in almost every direction, but principally from east to west; and filled with veins again broken, diversified by cross lines, and filled with other veins, and exhibiting marks of various successive phenomena of this kind.

Respecting the agents that produced the chasms in the primary strata, and the power by which they were filled with stony and metallic matter, it would be easy to speculate, but very difficult to reason, by legitimate philosophical induction.

The water-worn pebbles of chlorite schist, found cemented by oxide of tin, and of which an interesting account has been given by a member of the Society,* render it probable that the operation of water, either in the beds of rivers, or on the shores of lakes, or the ocean, preceded or accompanied the operations by which veins were produced and filled. All *crystallization* must be preceded by chemical solution,† or by a division of matter tantamount to *solution:* and elevation of temperature offers the most obvious means of explaining the production of combinations capable of depositing crystals.

Amongst the ancient lavas of Radicofani, I have seen crystals having the characters of felspar, imbedded in a black semi-crystallized mass, so as to constitute porphyry; and I am in possession of a specimen from Vesuvius, which I ascertained to belong to a stratum of lava, containing a cavity, in which felspar and mica

* [Mr. Carne, in Phil. Trans. for 1807.]

* [Is there not a crystallization, independent of solution; depending on the slow movement of the particles of matter in a long period of time, terminating in a crystallized arrangement? On the surface of ancient coins of bronze, such arrangements may be witnessed occasionally,—little elevated masses or protuberances containing well-formed crystals of copper, and protoxide of copper, in situations, and under circumstances—as in tombs and dry places—apparently incompatible with any kind of solution. I have recorded many such instances, in a paper entitled "Observations on the changes which have taken place in some ancient alloys of copper." (Phil. Trans. for 1826.) Does not a movement of particles take place in the annealing of glass, in the tempering of steel? If it can be established that the particles of a solid body, as steel or glass, even without chemical change, are capable of motion in relation to each other, why may it not be a property of matter generally, and be concerned in giving a crystallized character to rocks,—especially those of the secondary and transition class?]

have crystallized in their regular forms; and the separation and geometrical arrangement of their elements have evidently been the result of slow cooling.

The lavas of Languedoc, of the Vivarais, of Andernach, and of the Siennese and Roman States, offer numerous instances of the formation of tufas, which consist of *separate* crystals, and which have assumed the columnar arrangement. Near Aix, in Provence, I have seen a dyke between strata of limestone, which, at its lower extremity, has the character of *basalt*, and it is arranged like a basaltic dyke, in regular horizontal prisms; but its upper edge has the character of amorphous lava: and where it has been decomposed, and worn to a great depth, by the operation of a torrent, it has all the characters of a primitive sienite, being composed of large crystals of hornblende and felspar.

It is amongst extinct volcanoes, the surfaces of which have been removed by the action of air and water, and in which the interior parts of strata of lavas are exposed, that the most instructive examples of the operation of slow cooling upon heated masses, are to be found. It is difficult to conceive that water could have been the solvent of the different granitic and porphyritic formations; for in this case, some combinations of water with the pure earth ought to be found in them. Quartz ought to exist in the state of hydrate; and wavellite, not corundum, ought to be the state of alumina in granite.

To suppose the primary rocks in general to have been produced by the slow cooling of a mass formed by the combustion of the metallic bases of the earths, appears to me the most reasonable hypothesis; yet aqueous fusion must not be entirely excluded from our geological views. In many cases of crystallization, even in volcanic countries, this cause operates. Thus, in Ischia, as well as in Iceland, siliceous tufas are formed from hot springs; and in the Lake Albula, in the Lake of Solfaterra, near Tivoli, crystals of calcareous spar and of sulphur, separate from water, impregnated with carbonic acid and hepatic gas; and large strata of calcareous rocks formed evidently in late times by water, impregnated with carbonic acid, exist in various parts of Europe. The Traventine marble (marmor Tiburtinum) is a production of this kind,—and it is of this species of stone, that the Coliseum at Rome, and the Cathedral of St. Peter, are built. It is likewise employed in the ancient temples of Pæstum; and it rivals in durability, if not in beauty, the primary marble of Paros and Carrara.

VIII.

ON A DEPOSIT FOUND IN THE WATERS OF THE BATHS OF LUCCA.*

THE waters of the baths of Lucca, where they are of highest temperature, as in the Bagnie Calde, eject a considerable quantity of a substance from which a deposit is formed of a brownish yellow colour. Having collected a portion of this substance, and submitted it to chemical analysis, I found that it is composed of oxide of iron and of silica. The quantity of oxide in relation to that of the silica appeared to me to be as 4 to 3. I ought, however, to remark, that this conclusion was made from one experiment only, and that the balance at my disposal was imperfect, precluding much precision.

It is extremely probable that the silica and oxide of iron were dissolved together in the water and were simultaneously deposited. Indeed, when the silica is separated from the oxide of iron, by means of a weak acid, it has a gelatinous appearance; and it may be added, that the deposit in its natural state appears to be of uniform composition, even when examined with a magnifying glass.

The oxide of iron which I obtained from the waters was the peroxide; it is, however, very probable, that it existed in the state of protoxide, and was converted

[* This paper was originally communicated to the Academy of Sciences of Naples, and was republished in 1821, in the Annales de Chimie et de Physique, tom. xix; from which the translation above given has been made.]

into peroxide by the action of the air dissolved. One circumstance which seems to corroborate this opinion, is, that the colour of the water is not changed either by the addition of the triple prussiate of potassa or of infusion of galls.

The analogy which I have already established in my researches on the composition of the alkalies and the earths, between the base of silica and that of the boracic acid, and the facts brought forward by Mr. Smithson and M. Berzelius, tend to show the propriety of arranging silica in the class of acids. It appears probable, then, that the silica and oxide of iron existed in the hot water in chemical union, and that they separated in consequence of the cooling of the fluid on approaching the surface.

When the deposit is obtained direct from the water, by keeping it at rest in a clear vessel, it is manifest that the substance consists only of oxide of iron and of silica. When it is procured from the bottom of the baths, it is found mixed with carbonate of lime and with sand, both of which are evidently foreign. From a large number of experiments which I have made, I am satisfied that the water deposits nothing after leaving its source; but it appears certain, that the water which at its exit has a temperature of 112° F. must be hotter in the interior of the mountain, and consequently have a greater solvent power.

M. Battista Tassandori has found by evaporating a large quantity of these waters, that a deposit is obtained of oxide of iron and of silica. I find, on trial, that these two substances exist in it in the same proportions as in the native sediment.

In the Bath waters, a minute quantity of oxide of iron is found which is accompanied by silica. This earth, there is every reason to believe, is in many instances, the cause of oxide of iron being held in water in solution. Conjointly these facts lead to a probable theory of the formation of ochres. As regards the effects of the combinations of oxide of iron, and of silica on the animal economy, they belong entirely to medical science, and can be determined only by well directed trials, and by long experience.

IX.

ON THE STATE OF WATER AND AËRIFORM MATTER IN CAVITIES FOUND IN CERTAIN CRYSTALS.*

THERE are few inquiries in natural science more calculated to awaken our curiosity, than those relating to the changes which the matter composing the surface of our globe has undergone. The imagination is excited by the magnitude of the operations, by the obscurity of the phenomena, and the remoteness of the time at which they occurred; and all the intellectual powers are required to be brought into activity to find facts or analogies, or to institute experiments, by which they may be referred to known causes.

The crystallizations constituting the whole of the rocks which are usually called primary, and those found in such abundance, even in the rocks which are termed secondary, prove that a considerable part of the materials of the surface of the globe must have been either fluid or aëriform; for these are the only states from which the regular arrangements of the molecules of bodies constituting crystals can be produced.

Geologists are generally agreed, that the greater number of the crystalline mineral substances must have been previously in a liquid state; but different schools have supposed different causes for their solu-

[* From Phil. Trans. for 1822. Read before the Royal Society, June 13, 1822.]

tion; some attributing this effect principally to the agency of water, others to that of heat.

When, however, it is considered that the solvent power of water depends upon its temperature, and its deposition of solid matters upon its change of state or of temperature, and that being a gravitating substance, the same quantity must always belong to the globe, it becomes difficult to allow much weight to the arguments of the Wernerians or Neptunists, who have generally neglected, in their speculations, the laws of chemical attraction.

There are many circumstances, on the contrary, favourable to that part of the views of the Huttonians or Plutonists, relating to the cause of crystallization; such as the form of the earth, that of an oblate spheroid flattened at the poles; the facility with which heat, being a radiating substance, may be lost and dissipated in free space; and the observations which seem to show the present existence of a high temperature in the interior of the globe.

I have often, in the course of my chemical researches, looked for facts or experiments, which might throw some light on this interesting subject, but without success, till about three years ago; when, in considering the state of the fluid and aëriform matters included in certain crystals, it appeared to me, that these curious phenomena might be examined in a manner to afford some important arguments as to the causes of the formation of the crystal.

It is well known that water, and all fluids at usual temperatures, are more expansible by heat than siliceous or other earthy matters; and supposing these crystals to have been formed, and the water or fluid enclosed in them, at a pressure and temperatue not very unlike those of our existing atmosphere, this fluid ought to fill nearly the same space as when included, and the elastic fluid confined with it, supposing it non-absorbable, ought to be in the same state of density. On the contrary, if the earthy matter and the fluid separated from each other under a much higher temperature than that now belonging to the surface, a certain vacuum might be expected in the cavity from the contraction of the fluid, and if any gas were present, a considerable rarefaction of it; and though, supposing a much higher temperature on the surface of the globe, the atmosphere formed by aqueous vapour must have had much greater absolute weight, which, as liquids are *compressible*, must have influenced the volume of the fluid at the time it was inclosed, a circumstance which would render it impossible to draw any conclusion as to the exact temperature, yet still the experiments appeared to offer on any view, interesting results; and I was the more desirous of performing them, as I believe the nature of the fluid and aëriform matters included in rock crystals and other siliceous stones, has never been accurately ascertained.

Having purchased some crystals, and having others committed to my care by the liberality of my Brother Trustees of the British Museum, and of my friend Professor Buckland, I proceeded to make the necessary experiments upon them. It will be improper for me to take up the time of the Society by a minute description of my manipulations. Holes were drilled in the crystals by the use of diamonds, generally by Mr. Newman, under distilled water, or mercury, the gas was expelled by the introduction of wires, and the fluids included in the cavities were drawn out by means of fine capillary tubes, and experiments were afterwards made to determine the space they occupied, which had been accurately measured and marked upon the crystals. The chemical nature of the fluid and gas was determined by processes which were necessarily difficult from the smallness of the quantities operated upon; but which are too well known to the chemical philosophers of this Society to need description.

The first three crystals that I examined, were from Schemnitz, in Hungary; the cavities that they contained were proved not to be permeable to the atmosphere, by exposure to rarefied air, alone, and under water, in the receiver of an air-pump, a circumstance which it was necessary always to attend to, in order to render the experiment availing.

A cavity in one of the crystals was pierced under oil, one under distilled water, and one under mercury. In all of them the fluid rushed in when the cavity was opened, and the globule of elastic fluid contracted so as to appear from six to ten times less than before the experiment. The fluid in all the crystals (in two it was minutely examined,) was found to be water nearly pure, containing only a minute portion of the alkaline sulphates. The elastic fluid, as well as I could ascertain from the very minute quantities I could procure, appeared to be azote, unmixed with any other substance.

The largest cavity, which was in the crystal put into my hands by Professor Buckland, contained a space equal to 74·5 grains of mercury; the water in it equalled in volume 48·1 grain measures of mercury; and the globule of air, after the experiment, equalled in diameter a globule of mercury weighing 4·2 grains, so that the elastic fluid had contracted at least between six and seven times.

In the other experiments, the cavities being much smaller, the quantities of air and fluid could not be accurately measured; but there seemed to be nearly the same relation between the space filled by fluid, and that containing aëriform matter; and in all of them the contraction of the globule of aëriform matter was evidently greater, and in one instance to less than $\frac{1}{10}$ of its original bulk.

The fourth crystal that I experimented upon was of unknown locality; but I have reason to believe that it was from Guanaxuato, in Mexico, as it strongly resembled some that Mr. Heuland showed me from that place. The cavity in it was extremely small, and when pierced into, under distilled water, the globule of gas from being $\frac{1}{8}$ of an inch in diameter,* diminished so as to be less than $\frac{1}{25}$; so that its rarefaction was much greater in this than in the other instances; the water was too small in quantity to be minutely examined; it seemed to be nearly pure, producing a cloudiness barely perceptible in solutions of nitrate of silver and muriate of baryta.

It was an interesting point to ascertain whether the same circumstances occurred in productions found in rocks which have been generally considered as of igneous origin, such as the basaltic rocks in the neighbourhood of Vicenza, the chalcedonies of which so often afford included water. I found it much more easy to make experiments of this kind, and to procure specimens, which were abundantly supplied to me from the same sources as those I have just referred to; and

* I have not thought it necessary to refer to the heights of the barometer and thermometer in these experiments, as it is impossible to gain other than general results, upon quantities in which differences arising from atmospheric temperatures and pressure, would be quite unappreciable.

although some of these specimens proved to be permeable to the atmosphere, and to have been filled with water artificially, yet many occurred in which the sides of the cavity were absolutely impervious to air or water.

The results that I obtained were very analogous. Water containing very minute quantities of saline impregnations, occasioning barely a visible cloudiness in solutions of silver and of muriate of baryta, was found to be the fluid; the gas was azote, but it was in a much more rarefied state than in the rock crystals, being between sixty and seventy times as rare as atmospheric air.

The quantity of water was to the void space in greater proportion than in the rock-crystals. In the instance in which the most accurate experiment was made, namely, on the great specimen preserved in the collection of the British Museum, and which weighed 380 grains, the quantity of water was 29·9 grains, the space occupied by aëriform matter was equal to 11·7 grains of water, the volume of the globule of gas at the common pressure was to that of its rarefied volume as 1 to 63.

It occurred to me that atmospheric air might have been originally the elastic fluid included in these siliceous stones and in the crystals, and that the oxygen might have been separated from the azote by the attraction of the water, and a direct experiment seemed to confirm this idea. A chalcedony which had been bored was placed in water free from air under a receiver, which was exhausted till a portion of gas from the interior of the crystal had escaped into a proper receptacle. This gas examined by nitrous gas, was found to contain nearly as much oxygen as atmospheric air;

so that there is every reason to believe that the water had emitted oxygen during the exhaustion.

I endeavoured to find some calcareous secondary rocks, or crystals belonging to them, containing cavities, on which experiments of the same kind might be made; but in a number of trials I have as yet found none impermeable to the atmosphere; and the cavities of such, when bored, are always found to contain atmospheric air in a common state of density.

I was surprised to find that this was the case even with cavities in calcareous spar in the centre of limestone rock; yet these cavities which contain atmospheric air did not fill with water when the stone was placed in water under an exhausted receiver. When, however, it was dry, and placed in a receiver alternately exhausted and filled with hydrogen, the air that was produced by piercing the cavities, was found mixed with hydrogen; proving that the substance of the stone was permeable to elastic fluid. I hope soon to be able to make further researches on this subject; but in reasoning upon the vacuum, or rarefied state of the aëriform matter in the cavities of these rock crystals and chalcedonies, it appears difficult to account for the phenomenon, except on the supposition of their being formed at a higher temperature than that now belonging to the surface of the globe; and the most probable supposition seems to be, that the water and the silica were in chemical union, and separated from each other by cooling.

Water in the temperature of the arctic winter is constantly a crystallized body. As a fluid, its solvent powers are increased as its heat becomes higher, and when elastic, the density of its vapour is exalted in proportion to its heat; so that an atmosphere of steam, supplied from an indefinite source above water, would render it capable of receiving a very high degree of heat. Lime retains water in combination at a heat above 250° Fahrenheit; baryta retains it (even under ordinary pressure) at a strong red heat, and fuses with it. It is extremely likely that a liquid hydrate of silica would exist, under pressure, at high temperatures; and like all liquid bodies in the atmosphere, would probably contain small quantities of atmospheric air; and such a supposition only is necessary to account for the phenomenon presented by the water in rock crystal and chalcedony.

As, however, steam or aqueous vapour may be considered as having a share in these results, if it be supposed included in the cavity, no exact conclusions can be drawn from the apparent degree of contraction of the water; particularly as the late ingenious researches of Mr. Perkins show, that water is much more compressible than was formerly imagined; and the volume of water, however high its temperature, must be influenced by the pressure to which it is exposed; so that a certain compressing weight may not only impede, but altogether counteract the expansive force of heat.

Many speculations might be indulged in on this subject, but I shall not at present enter upon them; and I shall conclude by observing, that a fact, which has been considered by the Neptunists, above all others, as hostile to the idea of the igneous origin of crystalline rocks, namely, the existence of water in them, seems to afford a decisive argument in favour of the opinion it has been brought forward to oppose.

APPENDIX.

Since the foregoing pages were communicated to the Royal Society, I have made some new experiments on the same subject; all of them, except two, offered results of the same kind as these I have detailed, and upon such I shall not enter: but these two, from their peculiarity, will not, I trust, be thought unworthy of a particular notice,

In examining, with Mr. Heuland, the beautiful specimens of rock-crystals in the collection of Charles Hampden Turner, Esq., I observed one crystal which, Mr. Heuland informed me, was from La Gardette, in Dauphiné, that contained a considerable cavity, in which there was a viscid brownish liquid, resembling in its appearance and consistence linseed oil. As the void space or cavity filled with aëriform matter appeared considerable in proportion to the fluid, I expressed a desire to pierce the crystal; and Mr. Turner hearing of my wish, was so kind as to gratify it in the most polite and liberal manner, by presenting to me the specimen. With Mr. Newman's assistance, I made the usual experiments upon it. The cavity was pyramidal, and nearly the third of an inch in diameter. I soon ascertained that the fluid was not water, as it congealed and became opaque at a temperature of 56°. When the crystal was pierced under distilled water, the water rushed in and entirely filled the cavity, so that no other aëriform matter but the vapour of the substance could have been present; the water was rendered white and cloudy, apparently by the substance. I endeavoured to collect some of it for chemical examination, but it was too small in quantity (not equalling in volume $\frac{1}{6}$ of the volume of the cavity) to be submitted to analysis. It swam on the water, had no distinct taste, but a smell resembling naphtha; a portion of it taken out mixed with the water, when exposed to heat, acted like fixed oil, and it seemed to have a high temperature of ebullition. It inflamed, producing a white smoke.

The fact, of almost a perfect vacuum existing in a cavity containing an expansible but difficultly volatile substance, may be considered as highly favourable to the theory of the igneous origin of crystals: the other experiment is of a nature entirely different, though its result *may* be explained on the same supposition.

In examining a crystal in the collection of the Royal Institution, and which from its characters I believe to be from Capaó d'Olanda, Province of Minas Geraes, Brazil, I observed that the quantity of aëriform matter was unusually small in proportion to the quantity of fluid, in two or three cavities not occupying $\frac{1}{10}$ or $\frac{1}{12}$ of the space; and from the peculiarity of its motion, it appeared to be more likely to be compressed, than rarefied elastic fluid; and in piercing the sides of the cavity, I found that this was the case; it enlarged in volume from ten to twelve times; the fluid was water, but the gas was too minute in quantity to be examined.

It will be interesting to ascertain under what circumstances, and in what situations crystals of this kind are found. If they be supposed of igneous origin, they must have been formed under an immense weight of atmosphere or fluid, sufficient to produce a compression much more than adequate to compensate for the expansive effects of heat, a supposition which, in consequence of Mr. Perkin's experiments, already alluded to, may be easily formed.

X.

ON THE MAGNETIC PHENOMENA PRODUCED BY ELECTRICITY. IN A LETTER TO W. H. WOLLASTON, M.D. F.R.S.*

My dear Sir,

The similarity of the laws of electrical and magnetic attraction has often impressed philosophers; and many years ago in the progress of the discoveries made with the voltaic pile, some inquirers (particularly M. Ritter†) attempted to establish the existence of an identity or intimate relation between these two powers; but their views

* [From the Phil. Trans. for 1821. Read before the Royal Society, November 16, 1820.]

† M. Ritter asserted that a needle, composed of silver and zinc arranged itself in the magnetic meridian, and was slightly attracted and repelled by the poles of a magnet; and that a metallic wire, after being exposed in the voltaic circuit, took a direction N.E. and S.E. His ideas are so obscure that it is often difficult to understand them; but he seems to have had some vague notion that electrical combinations, when not exhibiting their electrical tension, were in a magnetic state, and that there was a kind of electro-magnetic meridian depending upon the electricity of the earth. See Annales de Chimie, t. lxiv. p. 80. Since this letter has been written, Dr. Marcet has been so good as to send me from Genoa, some pages of Aldini on Galvanism, and of Izarn's Manual of Galvanism, published at Paris more than sixteen years ago. M. Mojon, senior, of Genoa, is quoted in these pages as having rendered a steel needle magnetic, by placing it in a voltaic circuit for a great length of time. This, however, seems to have been dependent merely upon its place in the magnetic meridian, or upon an accidental curvature of it; but M. Romagnesi of Trente, is stated to have discovered that the pile of Volta caused a declination of the needle; the details are not given, but if the general statement be correct, the author could not have observed the same fact as M. Œrsted, but merely supposed that the needle has its magnetic poles altered after being placed in the voltaic circuit as a part of the electrical combination.

being generally obscure, or their experiments inaccurate, they were neglected; the chemical and electrical phenomena exhibited by the wonderful combination of Volta, at that time almost entirely absorbed the attention of scientific men; and the discovery of the fact of the true connection between electricity and magnetism, seems to have been reserved for M. Œrsted, and for the present year.

This discovery, from its importance, and unexpected nature, cannot fail to awaken a strong interest in the scientific world; and it opens a new field of inquiry into which many experimenters will undoubtedly enter: and where there are so many objects of research obvious, it is scarcely possible that similar facts should not be observed by different persons. The progress of science is, however, always promoted by a speedy publication of experiments; hence, though it is probable that the phenomena which I have observed may have been discovered before, or at the same time, in other parts of Europe, yet I shall not hesitate to communicate them to you, and through you to the Royal Society.

I found, in repeating the experiments of M. Œrsted with a voltaic apparatus of one hundred pairs of plates of four inches, that the south pole of a common magnetic needle (suspended in the usual way) placed under the communicating wire of platinum (the positive end of the apparatus being on the right hand) was strongly attracted by the wire, and remained in contact with it, so as entirely to alter the direction of the needle, and to overcome the magnetism of the earth. This I could only explain by supposing that the wire itself became magnetic during the passage of the electricity, through it, and direct experiments which I immediately made proved that this was the case. I threw some iron filings on a paper, and brought them near the communicating wire, when immediately they were attracted by the wire, and adhered to it in considerable quantities, forming a mass round it ten or twelve times the thickness of the wire: on breaking the communication, they instantly fell off, proving that the magnetic effect depended entirely on the passage of the electricity through the wire. I tried the same experiment on different parts of the wire, which was seven or eight feet in length, and about the twentieth of an inch in diameter, and I found that the iron filings were everywhere attracted by it; and making the communication with wires between different parts of the battery, I found that iron filings were attracted, and the magnetic needle affected in every part of the circuit.

It was easy to imagine that such magnetic effects could not be exhibited by the electrified wire without being capable of permanent communication to steel. I fastened several steel needles, in different directions, by fine silver wire to a wire of the same metal, of about the thirtieth of an inch in thickness, and eleven inches long, some parallel, others transverse, above and below in different directions: and I placed them in the electrical circuit of a battery of thirty pairs of plates of nine inches by five, and tried their magnetism by means of iron filings: they were all magnetic; those which were parallel to the wire attracted filings in the same way as the wire itself, but these in transverse directions, exhibited each two poles, which being examined by the test of delicate magnets, it was found that all the needles that were placed under the wire (the positive end of the battery being east) had their north poles on the south side of the wire, and their south poles on the north side; and that those placed over, had their south poles turned to the south, and their north poles turned to the north; and this was the case whatever was the inclination of the needles to the horizon. On breaking the connections all the steel needles that were on the wire in a transverse direction, retained their magnetism, which was as powerful as ever; whilst those which were parallel to the silver wire appeared to lose it at the same time as the wire itself.

I attached small longitudinal portions of wires of platinum, silver, tin, iron, and steel in transverse directions, to a wire of platinum that was placed in the circuit of the same battery. The steel and the iron wire immediately acquired poles in the same manner as in the last experiments; the other wires seemed to have no effect, except in acting merely as parts of the electrical circuit; the steel retained its magnetism as powerfully after the circuit was broken as before; the iron wire immediately lost a part of its polarity, and in a very short time the whole of it.

The battery was placed in different directions as to the poles of the earth; but the effect was uniformly the same. All needles placed transversely under the communicating wires, the positive end being on the right hand, had their north poles turned towards the face of the operator, and those above the wire their south poles; and on turning the wire round to the other side of the battery, it being in a longitudinal direction, and marking the side of the wire, the same side was always found to possess the same magnetism; so that in all arrangements of needles transversely round the wire, all the needles above had north and south poles opposite to those below, and those arranged vertically on one side, opposite to those arranged vertically on the other side.

I found that contact of the steel needles was not

necessary, and that the effect was produced instantaneously by the mere juxta-position of the needle in a transverse direction, and that through very thick plates of glass: and a needle that had been placed in a transverse direction to the wire, merely for an instant, was found as powerful a magnet as one that had been long in communication with it.

I placed some silver wire of one-twentieth of an inch, and some of one-fiftieth, in different parts of the voltaic circuit when it was completed, and shook some steel filings on a glass plate above them; the steel filings arranged themselves in right lines always at right angles to the axis of the wire; the effect was observed, though feebly, at the distance of a quarter of an inch above the thin wire, and the arrangement in lines was nearly to the same length on each side of the wire.

I ascertained, by several experiments, that the effect was proportional to the quantity of electricity passing through a given space, without any relation to the metal transmitting it; thus, the finer the wires the stronger their magnetism.

A zinc plate of a foot long and six inches wide, arranged with a copper plate on each side, was connected by a very fine wire of platinum, according to your method; and the plates were plunged an inch deep in dilute nitric acid. The wire did not sensibly attract fine steel filings. When they were plunged two inches, the effect was sensible; and it increased with the quantity of immersion. Two arrangements of this kind acted more powerfully than one; but when the two were combined, so as to make the zinc and copper plates the parts of one combination, the effect was very much greater. This was shown still more distinctly in the following experiment. Sixty zinc plates with double copper plates were arranged in alternate order, and the quantity of iron filings which a wire of a determinate thickness took up observed; the wire remaining the same, they were arranged so as to make a series of thirty; the magnetic effect appeared more than twice as great, that is, the wire raised more than double the quantity of iron filings.

The magnetism produced by voltaic electricity seems (the wire transmitting it remaining the same) exactly in the same ratio as the heat; and however great the heat of a wire, its magnetic powers were not impaired. This was distinctly shown in transmitting the electricity of twelve batteries of ten plates each of zinc, with double copper arranged as three, through fine platinum wire, which when so intensely ignited as to be near the point of fusion, exhibited the strongest magnetic effects, and attracted large quantities of iron filings and even small steel needles from a considerable distance.

As the discharge of a considerable quantity of electricity through a wire seemed necessary to produce magnetism, it appeared probable that a wire electrified by the common machine would not occasion a sensible effect; and this I found was the case, on placing very small needles across a fine wire connected with a prime conductor of a powerful machine and the earth. But as a momentary exposure in a powerful electrical circuit was sufficient to give permanent polarity to steel, it appeared equally obvious, that needles placed transversely to a wire at the time that the electricity of a common Leyden battery was discharged through it, ought to become magnetic; and this I found was actually the case, and according to precisely the same laws as in the voltaic circuit; the needle *under* the wire, the positive conductor being on the right hand, offering its north pole to the face of the operator, and the needle *above*, exhibiting the opposite polarity.

So powerful was the magnetism produced by the discharge of an electrical battery of 17 square feet highly charged, through a silver wire of $\frac{1}{20}$ of an inch, that it rendered bars of steel of two inches long and from $\frac{1}{20}$ to $\frac{1}{10}$ in thickness, so magnetic, as to enable them to attract small pieces of steel wire or needles; and the effect was communicated to a distance of five inches above or below or laterally from the wire, through water, or thick plates of glass, or metal electrically insulated.

The facility with which experiments were made with the common Leyden battery, enabled me to ascertain several circumstances which were easy to imagine, such as that a tube filled with sulphuric acid of $\frac{1}{4}$ of an inch in diameter, did not transmit sufficient electricity to render steel magnetic; that a needle placed transverse to the explosion through air, was less magnetized than when the electricity was passed through wire; that steel bars exhibited no polarity (at least at their extremities) when the discharge was made through them as part of the circuit, or when they were placed parallel to the discharging wire; that two bars of steel fastened together, and having the discharging wire placed through their common centre of gravity, showed little or no signs of magnetism after the discharge till they were separated, when they exhibited their north and south poles opposite to each other, according to the law of position.

These experiments distinctly showed that magnetism was produced wherever concentrated electricity passed through space; but the precise circumstances or law of its production were not obvious from them. When a magnet is made to act on steel filings, these filings arrange themselves in curves round the poles, but diverge in right lines; and in their adherence to each other form right lines, appearing as spicula. In the attraction of the filings round the wire in the voltaic circuit, on the contrary, they form one coherent mass, which would probably be perfectly cylindrical were it not for the influence of gravity. In first considering the subject, it appeared to me that there must be as many double poles as there could be imagined points of contact round the wire; but when I found the N. and S. poles of a needle uniformly attracted by the same quarters of the wire, it appeared to me that there must be four principal poles corresponding to these four quarters. You, however, pointed out to me that there was nothing definite in the poles, and mentioned your idea that the phenomena might be explained, by supposing a kind of revolution of magnetism round the axis of the wire, depending for its direction upon the position of the negative and positive sides of the electrical apparatus.

To gain some light upon this matter, and to ascertain correctly the relations of the north and south poles of steel magnetized by electricity to the positive and negative state, I placed short steel needles round a circle made on pasteboard, of about two inches and a half in diameter, bringing them near each other, though not in contact, and fastening them to the pasteboard by thread, so that they formed the sides of a hexagon inscribed within the circle. A wire was fixed in the centre of this circle, so that the circle was parallel to the horizon, and an electric shock was passed through the wire, its upper part being connected with the positive side of a battery, and its lower part with the negative. After the shock, all the wires were found magnetic, and each had two poles; the south pole being opposite to the north pole of the wire next to it, and *vice versa*; and when the north pole of a needle was touched with a wire, and that wire moved round the circle to the south pole of the same needle, its motion was opposite to that of the apparent motion of the sun.

A similar experiment was tried with six needles arranged in the same manner, with only this difference, that the wire positively electrified was below. In this case, the results were precisely the same, except that the poles were reversed; and any body, moved in the circle from the north to the south pole of the same needle, had its direction from east to west.

A number of needles were arranged as polygons in different circles round the same piece of paste-board, and made magnetic by electricity; and it was found, that in all of them, whatever was the direction of the paste-board, whether horizontal, or perpendicular, or inclined to the horizon, and whatever was the direction of the wire with respect to the magnetic meridian, the same law prevailed; for instance, when the positive wire was east, and a body was moved round the circle from the north to the south poles of the same wire, its motion (beginning with the lower part of the circle) was from north to south, or with the upper part from south to north: and when the needles were arranged round a cylinder of paste-board, so as to cross the wire, and a pencil mark drawn in the direction of the poles, it formed a spiral.

It was perfectly evident, from these experiments, that as many polar arrangements may be formed, as chords can be drawn in circles surrounding the wire; and so far these phenomena agree with your idea of revolving magnetism; but I shall quit this subject, which I hope you will yourself elucidate, for the information of the Society, to mention some other circumstances and facts belonging to the inquiry.

Supposing powerful electricity to be passed through two, three, four, or more wires, forming part of the same circuit, parallel to each other, in the same plane, or in different planes, it could hardly be doubted that each wire, and the space around it, would become magnetic in the same manner as a single wire, though in a less degree; and this I found was actually the case. When four wires of fine platinum were made to complete a powerful voltaic circuit, each wire exhibited its magnetism in the same manner, and steel filings on the sides of the wires opposite attracted each other.

As the filings on the opposite sides of the wire attracted each other, in consequence of their being in opposite magnetic states, it was evident, that if the similar sides could be brought in contact, steel filings upon them would repel each other. This was very easily tried with two voltaic batteries, arranged parallel to each other, so that the positive end of one was opposite to the negative end of the other: steel filings upon two wires of platinum, joining the extremities, strongly repelled each other. When the batteries were arranged in the *same* order, i.e. positive opposite to positive, they attracted each other; and wires of platinum (without filings), and fine steel wire (still more strongly), exhibited similar phenomena of attraction and repulsion under the same circumstances.

As bodies magnetized by electricity put a needle in motion, it was natural to infer, that a magnet would put bodies magnetized by electricity in motion: and this I found was the case. Some pieces of wire of platinum, silver, and copper, were placed separately upon two knife edges of platinum, connected with two ends of a powerful voltaic battery, and a magnet presented to them: they were all made to roll along the knife edges, being attracted when the north pole of the magnet was presented, the positive side of the battery being on the right hand, and repelled when it was on the left hand; and, *vice versa*, changing the pole of the magnet. Some folds of gold leaf were placed across the same apparatus, and the north pole of a powerful magnet held opposite to them: the folds approached the magnet, but did not adhere to it. On the south pole being presented, they receded from it.

I will not indulge myself by entering far into the theoretical parts of this subject; but a number of curious speculations cannot fail to present themselves to every philosophical mind, in consequence of the facts developed: such as, whether the magnetism of the earth may not be owing to its electricity, and the variation of the needle to the alterations in the electrical currents of the earth, in consequence of its motions, internal chemical changes, or its relation to solar heat; and whether the luminous effects of the auroras at the poles are not shown, by these new facts, to depend on electricity. This is evident, that if strong electrical currents be supposed to follow the apparent course of the sun, the magnetism of the earth ought to be such as it is found to be.

But I will quit conjectures, to point out a simple mode of making powerful magnets, namely, by fixing bars of steel across, or circular pieces of steel fitted for making horse-shoe magnets, round the electrical conductors of buildings in elevated and exposed situations.*

* There are many facts recorded in the Philosophical Transactions, which prove the magnetizing powers of lightning; one in particular,

The experiments detailed in these pages were made with the apparatus belonging to the Royal and London Institutions: and I was assisted in many of them by Mr. Pepys, Mr. Allen, and Mr. Stodart, and in all of them by Mr. Faraday.*

I am, my dear Sir,

Very sincerely yours,

HUMPHRY DAVY.

Lower Grosvenor Street,
Nov. 12, 1820.

where a stroke of lightning, passing through a box of knives, rendered most of them powerful magnets. See Phil. Trans. No. 157, p. 520; and No. 437, p. 57.

* All the experiments detailed in this paper, except those mentioned page 224, were made in the course of October, 1820; the last arose in consequence of a conversation with Dr. Wollaston, and were made in the beginning of November. I find, by the Annales de Chimie et de Physique, for September, which arrived in London, November 24, that M. Arago has anticipated me in the discovery of the attractive and magnetizing powers of the wires in the voltaic circuit; but the phenomena presented by the action of common electricity (which, I believe, as yet, have been observed by no other person), induce me still to submit my paper to the Council of the Royal Society. Before any notice arrived of the researches of the French philosophers, I had tried, with Messrs. Allen and Pepys, an experiment, which M. Arago likewise thought of,— whether the arc of flame of the voltaic battery would be affected by the magnet; but from the imperfection of our apparatus, the results were not decisive. I hope soon to be able to repeat it under new circumstances.

I have made various experiments, with the hope of affecting electrified wires by the magnetism of the earth, and of producing chemical changes by magnetism; but without any successful results.

Since I have perused M. Ampere's elaborate treatise on electro-magnetic phenomena, I have passed the electrical shock along a spiral wire, twisted round a glass tube, containing a bar of steel, and I found that the bar was rendered powerfully magnetic by the process.

Without meaning to offer any decided opinion on that gentleman's ingenious views, I shall beg permission to mention two circumstances, which seem to me unfavourable to the idea of the identity of electricity and magnetism;—1st, the great distances to which magnetism is communicated by common electricity, (I found that a steel bar was made magnetic at 14 inches distance from a wire transmitting an electric shock from about 70 feet of charged surface); and, 2nd, that the effect of magnetizing at a distance, by electricity, takes place with the same readiness through air and water, glass, mica, or metal; i. e. through conductors and non-conductors.

XI.

FARTHER RESEARCHES ON THE MAGNETIC PHENOMENA PRODUCED BY ELECTRICITY; WITH SOME NEW EXPERIMENTS ON THE PROPERTIES OF ELECTRIFIED BODIES IN THEIR RELATIONS TO CONDUCTING POWERS AND TEMPERATURE.*

I. In my letter to Dr. Wollaston on the new facts discovered by M. Œrsted, which the Society has done me the honour to publish, I mentioned that I was not able to render a bar of steel magnetic by transmitting the electrical discharge across it through a tube filled with sulphuric acid; and I have likewise mentioned, that the electrical discharge passed across a piece of steel through air, rendered it less magnetic than when passed through a metallic wire; and I attributed the first circumstance to the sulphuric acid being too bad a conductor to transmit a sufficient quantity of electricity for the effect; and the second, to the electricity passing through air in a more diffused state than through metals.

To gain some distinct knowledge on the relations of the different conductors to the magnetism produced by electricity, I instituted a series of experiments, which led to very decisive results, and confirmed my first views.

II. I found that the magnetic phenomena were precisely the same, whether the electricity was small in

* [From the Phil. Trans. for 1821. Read before the Royal Society, July 5, 1821.]

quantity, and passing through good conductors of considerable magnitude; or, whether the conductors were so imperfect as to convey only a small quantity of electricity; and in both cases they were neither attractive of each other, nor of iron filings, and not affected by the magnet; and the only proof of their being magnetic, was their occasioning a certain small deviation of the magnetized needle.

Thus a large piece of charcoal placed in the circuit of a very powerful battery, being a very bad conductor compared with the metals, would not affect the compass needle at all, unless it had a very large contact with the metallic part of the circuit; and if a small wire was made to touch it in the circuit only in a few points, that wire did not gain the power of attracting iron filings; though when it was made to touch a surface of platinum-foil coiled round the end of the charcoal, a slight effect of this kind was produced. And in a similar manner fused hydrat of potassa, one of the best of the imperfect conductors, could never be made to exert any attractive force on iron-filings, nor could the smallest filaments of cotton moistened by solution of hydrat of potassa, placed in the circuit, be made to move by the magnet; nor did steel needles floating on cork on an electrized solution of this kind placed in the voltaic circuit, gain any polarity; and the only proof of the magnetic power of electricity passing through such a fluid, was afforded by its effect upon the magnetized needle, when the metallic surfaces, plunged in the fluid, were of considerable extent. That the mobility of the parts of fluids, do not interfere with their magnetic powers as developed by electricity, I proved by electrifying mercury and Newton's metal fused, in small tubes. These tubes, placed in a proper voltaic circuit, attracted iron filings, and gave magnetic powers to needles; nor did any agitation of the mercury or metal within, either in consequence of mechanical motion or heat, alter or suspend their polarity.

III. Imperfect conducting fluids do not give polarity to steel when electricity is passed through them; but electricity passed through air produces this effect. Reasoning on this phenomenon, and on the extreme mobility of the particles of air, I concluded, as M. Arago had likewise done from other considerations, that the voltaic current in air would be affected by the magnet. I failed in my first trial, which I have referred to in a note to my former paper, and in other trials made since by using too weak a magnet; but I have lately had complete success; and the experiment exhibits a very striking phenomenon.

Mr. Pepys having had the goodness to charge the great battery of the London Institution, consisting of two thousand double plates of zinc and copper, with a mixture of 1168 parts of water, 108 parts of nitrous acid, and 25 parts of sulphuric acid, the poles were connected by charcoal, so as to make an arc, or column of electrical light, varying in length from one to four inches according to the state of rarefaction of the atmosphere in which it was produced; and a powerful magnet being presented to this arc or column, having its pole at a very acute angle to it, the arc or column was attracted or repelled with a rotatory motion, or made to revolve by placing the poles in different positions, according to the same law as the electrified cylinders of platinum described in my last paper, being repelled when the negative pole was on the right hand by the north pole of the magnet, and attracted by the south pole, and *vice versa*.

It was proved by several experiments that the motion depended entirely upon the magnetism, and not upon the electrical inductive power of the magnet; for masses of soft iron, or of other metals, produced no effect.

The electrical arc or column of flame was more easily affected by the magnet, and its motion was more rapid, when it passed through dense than through rarefied air; and in this case the conducting medium, or chain of aëriform particles was much shorter.

I tried to gain similar results with currents of common electricity sent through flame, and in vacuo. They were always affected by the magnet; but it was not possible to obtain so decided a result as with voltaic electricity, because the magnet itself became electrical by induction, and that whether it was insulated, or connected with the ground.*

IV. Metals, it is well known, readily transmit large quantities of electricity; and the obvious limit to the quantity which they are capable of transmitting seems to be their fusibility, or volatilization by the heat which electricity produces in its passage through bodies.

Now I had found in several experiments, that the intensity of this heat was connected with the nature of the medium by which the body was surrounded; thus a wire of platinum which was readily fused by transmitting the charge from a voltaic battery in the exhausted receiver of an air-pump, acquired in air a much lower degree of temperature. Reasoning on this

* I made several experiments on the effects of currents of electricity simultaneously passing through air in different states of rarefaction, in the same and different directions, both from the voltaic and common electrical batteries; but I could not establish the fact of their magnetic attractions or repulsions with regard to each other, which probably was owing to the impossibility of bringing them sufficiently near.

circumstance, it occurred to me, that by placing wires in a medium much denser than air, such as ether, alcohol, oils, or water, I might enable them to transmit a much higher charge of electricity than they could convey without being destroyed in air; and thus not only gain some new results as to the magnetic states of such wires, but likewise, perhaps, determine the actual limits to the powers of different bodies to conduct electricity, and the relations of these powers.

A wire of platinum of $\frac{1}{220}$, of three inches in length, was fused in air, by being made to transmit the electricity of two batteries of ten zinc plates of four inches with double copper, strongly charged: a similar wire was placed in sulphuric ether, and the charge transmitted through it. It became surrounded by globules of gas; but no other change took place; and in this situation it bore the discharge from twelve batteries of the same kind, exhibiting the same phenomenon. When only about an inch of it was heated by this high power in ether, it made the ether boil, and became white hot under the globules of vapour, and then rapidly decomposed the ether, but it did not fuse. When oil or water was substituted for the ether, the length of the wire remaining the same, it was partially covered with small globules of gas, but did not become red hot.

On trying the magnetic powers of this wire in water, they were found to be very great, and the quantity of iron filings that it attracted was such as to form a cylinder round it of nearly the tenth of an inch in diameter.

To ascertain whether short lengths of fine wire, prevented from fusing by being kept cool, transmitted the whole electricity of powerful voltaic batteries, I made a second independent circuit from the ends of the battery with silver wires in water, so that the chemical decomposition of the water indicated a residuum of electricity in the battery. Operating in this way, I found that an inch of wire of platinum of $\frac{1}{220}$, kept cool by water, left a great residual charge of electricity in a combination of twelve batteries, of the same kind as those above mentioned; and after making several trials, I found that it was barely adequate to discharge six batteries.

V. Having determined that there was a *limit* to the quantity of electricity which wires were capable of transmitting, it became easy to institute experiments on the different conducting powers of different metallic substances, and on the relation of this power to the temperature, mass, surface, or length of the conducting body, and to the conditions of electro-magnetic action.

These experiments were made as nearly as possible under the same circumstances, the same connecting copper-wires being used in all cases, their diameter being more than one-tenth of an inch, and the contact being always preserved perfect; and parts of the same solutions of acid and water were employed in the different batteries, and the same silver wires and broken circuit with water were employed in the different trials; and when no globules of gas were observed upon the negative silver wire of the second circuit, it was concluded that the metallic conducting chain, or the primary circuit, was adequate to the discharge of the combination. To describe more minutely all the precautions observed, would be tedious to those persons who are accustomed to experiments with the voltaic apparatus, and unintelligible to others; and after all, in researches of this nature, it is impossible to gain more than approximations to true results; for the gas disengaged upon the plates, the different distances in the connecting plates, and the slight difference of time in making the connections, all interfere with their perfect accuracy.

The most remarkable general result that I obtained by these researches, and which I shall mention first, as it influences all the others, was, that *the conducting power of metallic bodies varied with the temperature, and was lower, in some inverse ratio, as the temperature was higher.*

Thus a wire of platinum of $\frac{1}{220}$, and three inches in length, when kept cool by oil, discharged the electricity of two batteries, or of twenty double plates: but when suffered to be heated by exposure in the air, it barely discharged one battery.

Whether the heat was occasioned by the electricity, or applied to it from some other source, the effect was the same. Thus a wire of platinum, of such length and diameter, as to discharge a combination without being considerably heated; when the flame of a spirit-lamp was applied to it, so as to make a part of it red-hot, lost its power of discharging the whole electricity of the battery, as was shown by the disengagement of abundance of gas in the secondary circuit; which disengagement ceased as soon as the source of heat was withdrawn.

There are several modes of exhibiting this fact, so as to produce effects which, till they are witnessed, must almost appear impossible. Thus, let a fine wire of platinum, of four or five inches in length, be placed in the voltaic circuit, so that the electricity, passing through it, may heat the whole of it to redness, and let the flame of a spirit-lamp be applied to any part of it, so as to heat that part to whiteness, the rest of the wire will instantly become cold below the point of visible ignition. For the converse of the experiments, let a piece of ice, or a stream of cold air, be applied to a part of the wire; the other parts will immediately become much hotter; and from a red will rise to a white heat. The quantity of electricity that can pass through that part of the wire submitted to the changes of temperature is so much smaller when it is hot than when it is cold, that the absolute temperature of the whole wire is diminished by heating a part of it, and, *vice versa*, increased by cooling a part of it.

In comparing the conducting power of different metals, I found much greater differences than I expected. Thus six inches of silver wire of $\frac{1}{220}$ discharged the whole of the electricity of sixty-five pairs of plates of zinc and double copper, made active by a mixture of about one part of nitric acid of commerce, and fifteen parts of water. Six inches of copper wire of the same diameter discharged the electricity of fifty-six pairs of the same combination; six inches of tin of the same diameter, carried off that of twelve only; the same quantity of wire of platinum that of eleven, and of iron that of nine. Six inches of wire of lead of $\frac{1}{200}$, seemed equal in their conducting powers to the same length of copper wire of $\frac{1}{220}$. All the wires were kept as cool as possible, by immersion in a basin of water.*

I made a number of experiments of the same kind, but the results were never precisely alike, though they sometimes approached very near each other. When the batteries were highly charged, so that the intensity of the electricity was higher, the differences were less be-

* Water is so bad a conductor, that in experiments of this kind, its effects may be neglected altogether; and these effects were equal in all the experiments.

tween the best and worst conductors, and they were greater when the charge was extremely feeble. Thus, with a fresh charge of about one part of nitric acid and nine parts of water, wires of $\frac{1}{220}$ of silver and platinum, five inches long, discharged respectively the electricity of 30, and 7 double plates.

Finding that when different portions of the same wire plunged in a non-conducting fluid, were connected with different parts of the same battery equally charged, their conducting powers appeared in the inverse ratio of their lengths; so, when six inches of wire of platinum of $\frac{1}{220}$ discharged the electricity of 10 double plates, three inches discharged that of 20, $1\frac{1}{2}$ inch that of 40, and one inch that of 60; it occurred to me that the conducting powers of the different metals might be more easily compared in this way, as it would be possible to make the contacts in less time than when the batteries were charged, and consequently with less variation in the charge.

Operating in this way, I ascertained, that in discharging the electricity of 60 pairs of plates, one inch of platinum was equal to about 6 inches of silver, to $5\frac{1}{2}$ inches of copper, to 4 of gold, to 3·8 of lead, to about $\frac{9}{10}$ of palladium, and $\frac{8}{10}$ of iron, all the metals being in a cooling fluid medium.

I found, as might have been expected, that the conducting power of a wire for electricity, in batteries of the size and number of plates just described, was nearly directly as the mass; thus, when a certain length of wire of platinum discharged one battery,* the same length of wire of six times the weight discharged six batteries; and the effect was exactly the same, provided

* A foot of this wire weighed 1·13 grain; a foot of the other, 6·7 grains.

the wires were kept cool, whether the mass was a single wire, or composed of six of the smaller wires in contact with each other. This result alone showed, that surface had no relation to conducting power, at least for electricity of this kind, and it was more distinctly proved by a direct experiment. Equal lengths and equal weight of wire of platinum, one round, and one flattened by being passed transversely through rollers, so as to have six or seven times the surface, were compared as the conducting powers: the flattened wire was the best conductor in air, from its greater cooling powers, but in water no difference could be perceived between them.

VI. I tried to make a comparison between the conducting powers of fluid menstrua and charcoal, and those of metals. Six inches of platinum foil, an inch and one-fifth broad, were placed in a vessel which could be filled with any saline solution; and a similar piece of platinum placed opposite at an inch distance: the whole was then made part of a voltaic circuit, which had likewise another termination by silver wires in water; and solution of salts added, till gas ceased to be liberated from the negative silver wire. In several trials of this kind it was found that the whole of the surface of six inches, even with the strongest solutions of common salt, was insufficient to carry off the electricity even of two pair of plates; and a strong solution of potassa carried off the electricity of three pair of plates only; whereas an inch of wire of platinum of $\frac{1}{220}$ (as has been stated) carried off all the electricity of sixty pair of plates. The gas liberated upon the surface of the metals when they are placed in fluids, renders it impossible to gain accurate results; but the conducting power of the best fluid conductors, it seems probable from these experiments, must be some hundreds of thousand times less than those of the worst metallic conductors.

A piece of well-burnt compact box-wood charcoal was placed in the circuit, being three-tenths of an inch wide by one-tenth thick, and connected with large surfaces of platinum. It was found that one inch and two-tenths carried off the same quantity of electricity as six inches of wire of platinum of $\frac{1}{220}$.

VII. I made some experiments with the hope of ascertaining the exact change of ratio of the conducting powers dependent upon the change of the intensity and quantity of electricity; but I did not succeed in gaining any other than the general result, that the higher the intensity of the electricity, the less difficulty it had in passing through bad conductors; and several remarkable phenomena depend upon this circumstance.

Thus, in a battery where the quantity of the electricity is very great and the intensity very low, such as one composed of plates of zinc and copper, so arranged as to act only as single plates of from twenty to thirty feet of surface each, and charged by a weak mixture of acid and water, charcoal, made to touch only in a few points is almost as much an insulating body as water, and cannot be ignited, nor can wires of platinum be heated when their diameter is less than one-eightieth of an inch, and their length three or four feet; and a foot of platinum wire of one-thirtieth is scarcely heated by such a battery, whilst the same length of silver wire of the same diameter is made red-hot; and the same length of thicker wires of platinum or iron are intensely heated.

The heat produced where electricity of considerable intensity is passed through conductors, must always interfere with the exact knowledge of the charges of their conducting powers, as is proved by the following experiments. A battery of twenty pair of plates of zinc, and copper plates ten inches by six, was very highly charged with a mixture of nitric acid and water, so as to exhibit a considerable intensity of electrical action, and the relative conducting powers of silver and platinum in air and water ascertained by means of it. In air, six inches of platinum of one-eightieth, discharged only four double plates, whilst six inches of silver wire, of the same diameter, discharged the whole combination; the platinum was strongly ignited in this experiment, whilst the silver was scarcely warm to the touch. On cooling the platinum wire, by placing it in water, it was found to discharge ten double plates. When the intensity of the electricity is very high, however, even the cooling powers of fluid media are of little avail; thus I found that fine wire of platinum was fused by the discharge of a common electrical battery under water; so that the conducting power must always be diminished by the heat generated in a greater proportion as the intensity of the electricity is higher.

It might at first view be supposed, that when a conductor placed in the circuit left a residuum of electricity in any battery, increase of the power of the battery, or of its surface, would not enable it to carry through any additional quantity. This, however, is far from being the case. When saline solutions were placed in the circuit of a battery of twenty plates, though they discharged a very small quantity only of the electricity when the troughs were only one-fourth full, yet their chemical decomposition exhibited the fact of a much larger quantity passing through them, when the cells were filled with fluid.

A similar circumstance occurred with respect to a wire of platinum, of such a length as to leave a considerable residuum in a battery when only half its surface was used; yet when the whole surface was employed, it became much hotter, and nevertheless left a still more considerable residuum.

VIII. I found long ago, that in increasing the number of alternations of similar plates, the quantity of electricity seemed to increase as the number, at least as far as could be judged of by the effects of heat upon wires; but only within certain limits, beyond which the number appeared to diminish rather than increase the quantity. Thus the two thousand double plates of the London Institution, when arranged as one battery, would not ignite so much wire as a single battery of ten double plates with double copper.

It is not easy to explain this result. Does the intensity mark the rapidity of the motion of the electricity? or merely its diminished attraction for the matter on which it acts? and does this attraction become less in proportion as the circuit, through which it passes, or in which it is generated, contains a greater number of alternations of bad conductors?

Mr. Children, in his account of the experiments made with his battery of large plates, has ingeniously referred the heat produced by the passage of electricity through conductors, to the resistance it meets with, and has supposed, what proves to be the fact, that the heat is in some inverse ratio to the conducting power. The greatest heat, however, is produced in air, where there is reason to suppose the least resistance; and as the presence of heat renders bodies worse conductors, another view may be taken, namely, that the excitation of heat occasions the imperfection of the conducting power. But till the causes of heat and of electricity are known, and of that peculiar constitution of matter which excites the one, and transmits or propagates the other, our reasoning on this subject must be inconclusive.

I found that when equal portions of wires of the same diameter, but of different metals, were connected together in the circuit of a powerful voltaic battery, acting as two surfaces, the metals were heated in the following order:—iron most, then palladium, then platinum, then tin, then zinc, then gold, then lead, then copper, and silver least of all. And from one experiment, in which similar wires of platinum and silver joined in the same circuit were placed in equal portions of oil, it appeared that the generation of heat was nearly inversely as their conducting power. Thus the silver raised the temperature of the oil only four degrees, whilst the platinum raised it twenty-two. The same relations to heat seem to exist, whatever is the intensity of the electricity; thus, circuits of wires placed under water and acted on by the common electrical discharge, were heated in the same order as by the voltaic battery, as was shown by their relative fusion; thus, iron fusing before platinum, platinum before gold, and so on.

If a chain be made of wire of platinum and silver, in alternate links soldered together, the silver wire being four or five times the diameter of the platinum, and placed in a powerful voltaic circuit, the silver links are not sensibly heated, whilst all those of the platinum become intensely and equally ignited. This is an important experiment for investigating the nature of *heat.* If heat be supposed a substance, it cannot be imagined to be expelled from the platinum; because an unlimited quantity may be generated from the same platinum, *i.e.* as long as the electricity is excited, or as often as it is renewed. Or if it be supposed to be identical with, or an element of, electricity, it ought to bear some relation to its quantity, and might be expected to be the same in *every* part of the chain, or greatest in those parts nearest the battery.

IX. The magnetism produced by electricity, though with the same conductors it increases with the heat, as I mentioned in my last paper; yet with different conductors I find it follows a very different law. Thus, when a chain is made of different conducting wires, and they are placed in the same circuit, they all exhibit equal magnetic powers, and take up equal quantities of iron filings. So that the magnetism seems directly as the quantity of electricity which they transmit. And when, in a highly powerful voltaic battery, wires of the same diameters and lengths, but of which the best conducting is incapable of wholly discharging the battery, are made, separately and successively, to form the circuit, they take up different quantities of iron filings, in some direct proportion to their conducting powers.

Thus, in one experiment, two inches of wire of one-thirtieth of an inch being used, silver took up thirty-two grains, copper twenty-four, platinum eleven, and iron eight and two-tenths.

XII.

ON THE ELECTRICAL PHENOMENA EXHIBITED IN VACUO.*

THE production of heat and light by electrical discharges; the manner in which chemical attractions are produced, destroyed, or modified by changes in the electrical states of bodies; and the late important discovery of the connection of magnetism with electricity, have opened an extensive field of inquiry in physical science, and have rendered investigations, concerning the nature of electricity, and the laws by which it is governed, and the properties that it communicates to bodies, much more interesting than at any former period of the history of philosophy.

Is electricity a subtile elastic fluid?—or are electrical effects merely the exhibition of the attractive powers of the particles of bodies? Are heat and light elements of electricity, or merely the effects of its action? Is magnetism identical with electricity, or an independent agent, put into motion or activity by electricity? Queries of this kind might be considerably multiplied, and stated in more precise and various forms: the solution of them, it must be allowed, is of the highest importance; and though some persons have undertaken to answer them in the most positive manner, yet there are, I believe, few sagacious reasoners, who think that our

* [From the Phil. Trans. for 1822. Read before the Royal Society, December 20, 1821.]

present data are sufficient to enable us to decide on such very abstruse and difficult parts of corpuscular philosophy.

It appeared to me an object of considerable moment, and one intimately connected with all these queries,—*the relations of electricity to space, as nearly void of matter as it can be made on the surface of the earth;* and, in consequence, I undertook some experiments on the subject.

It is well known to the Fellows of this Society, who have considered the subject of electricity, that Mr. Walsh believed that the electrical light was not producible in a perfect torricellian vacuum; and that Mr. Morgan drew the same inference from his researches: and likewise concluded that such a vacuum prevented the charging of coated glass. Now, it is well known, that in the most perfect vacuum that can be made in the torricellian tube, vapour of mercury, though of extremely small density, exists; I could not help, therefore, entertaining a doubt as to the perfect accuracy of these results; and I resolved not only to examine them experimentally, but likewise, by using a comparatively fixed metal in fusion for making the vacuum, to exclude, as far as was possible, the presence of any volatile matter.

The apparatus that I employed was extremely simple —(Plate xi), and consisted of a curved glass tube, with one leg closed, and longer than the other. In this closed leg a wire of platinum was hermetically cemented, for the purpose of transmitting the electricity; or to ascertain the power of the vacuum to receive a charge, it was coated with foil of tin or platinum. The open end, when the closed leg had been filled with mercury or any other metal, was exhausted, either by being placed under the receiver, or connected with the stop-cock of an excellent air-pump; and, in some cases, to ensure greater accuracy, the exhaustion was made after the tube and apparatus had been filled with hydrogen.*

Operating in this way, it was easy to procure a vacuum, either of a large or small size,—for the rarefied air or gas could be made to balance a column of fluid metal of any length, from twenty inches to the twentieth of an inch; and by using only a small quantity of metal, it could be more easily purged of air.

I shall first mention the results I obtained with quicksilver. I found that, by using recently-distilled quicksilver in the tubes, and boiling it in vacuo six or seven times, from the top to the bottom, and from the bottom to the top, making it vibrate repeatedly, by striking it with a small piece of wood, a column was obtained in the tube, free from the smallest particle of air; but a phenomenon occurred, in discovering the cause of which, I had a great deal of trouble. When I used a short tube of four or five inches long only, I found, that after continued boiling and much agitation of the mercury, though there was no appearance of elastic matter, when the mercury adhered strongly in the upper part of the tube, yet that, after electrization, or even on suffering the mercury to pass slowly back into the closed part, a minute globular space sometimes appeared: I thought at first this was air, which, though so highly rarefied, as it must have been by the exhaustion, adhered to the mercury; and I endeavoured by long boiling the mercury in an exhausted *double* syphon, and making the vacuum in one of the curves, to prevent entirely the presence of air: but the phenomenon always occurred when there was no strong adhesion of the mercury to the glass. This, and another circumstance

* The figure will best explain the form of the apparatus.

—namely, that when the leg in which the torricellian vacuum was made, was fifteen or sixteen inches long, the phenomenon was very rarely perceptible, and always disappeared when the tube was inverted, and the mercury made to strike the top with some force,—led me to conclude that the minute space was really filled with the vapour of mercury; the attraction of the particles of the fluid mercury for each other preventing their actual contact with the glass, except when this contact was forcibly made by mechanical means; and I soon proved that this was the case: for, by causing the mercury, when its column was short, to descend into the more perfect, from the less perfect vacuum, with more or less velocity, I could make the space more or less, or cause its disappearance altogether, in which last case the cohesion between the mercury and the glass was always extremely strong.

I found that in all cases when the mercurial vacuum was perfect, it was permeable to electricity, and was rendered luminous by either the common spark, or the shock from a Leyden jar, and the coated glass surrounding it became charged; but the degree of intensity of these phenomena depended upon the temperature: when the tube was very hot, the electric light appeared in the vapour of a bright green colour, and of great density; as the temperature diminished, it lost its vividness; and when it was artificially cooled to 20° *below* zero of Fahrenheit, it was so faint, as to require considerable darkness to be perceptible.

The charge likewise communicated to the tin or platinum foil was higher, the higher the temperature: which, like the other phenomena, must depend upon the different density of the vapour of mercury; and at 0 Fahrenheit, it was very feeble indeed.

A very beautiful phenomenon occurred in boiling the mercury in the exhausted tube, which showed the great brilliancy of the electrical light in pure dense vapour of mercury. In the formation and condensation of the globules of mercurial vapour, the electricity produced by the friction of the mercury against the glass, was discharged through the vapour, with sparks so bright as to be visible in day-light.

In all cases when the minutest quantity of rare air was introduced into the mercurial vacuum, the colour of the light produced by the passage of the electricity, changed from green to sea-green; and by increasing the quantity, to blue and purple; and when the temperature was low, the vacuum became a much better conductor.

I tried to get rid of a portion of the mercurial vapour, by using a difficultly fusible amalgam of mercury and tin, which was made to crystallize, by cooling in the tube; but the results were precisely the same as when pure mercury was used.

I tried to make a vacuum above the fusible alloy of bismuth; but I found it so liable to oxidate and dirt the tube, that I soon renounced further attempts of this kind.

On a vacuum above fused tin, I made a number of experiments; and by using freshly-cut pieces of grain tin, and fusing them in a tube void, after being filled with hydrogen, and by long-continued heat and agitation, I had a column of fused tin, which appeared entirely free from gas; yet the vacuum made above this exhibited the same phenomena as the mercurial vacuum. At temperatures below 0°, the light was yellow, and of the palest phosphorescent kind, requiring almost absolute darkness to be perceived; and it was not perceptibly increased by heat.

I made two experiments on electrical and magnetic repulsions and attractions in the mercurial vacuum, by attaching to the platinum wire two fine wires, in one case of platinum, in the other of steel, terminated by minute spherules of the same metals: I found that they repelled each other, when the wire was electrified in the most perfect mercurial vacuum, as they would have done in the usual cases; and the steel globules were as obedient to the magnet as in the air; which last result it was easy to anticipitate.

In some of the first of these experiments, I used a wire for connecting the metal with the stop-cock; but latterly, the rarefied air or gas was the only chain of communication; and this circumstance enabled me to ascertain that the feebleness of the light in the more perfect vacuum, was not owing merely to a smaller quantity of electricity passing through it; for the same discharge which produced a faint green light in the upper part of the tube, produced a bright purple light in the lower part, and a strong spark in the atmosphere.

The boiling point of pure olive oil is not much below that of mercury; and the butter or chloride of antimony (antimonane) boils at about 388° Fahrenheit. I tried both these substances in the vacuum, and found, as might be expected, that the light produced by the electricity passing through the vapour of the chloride was much more brilliant than that produced by it in passing through the vapour of the oil; and in the last, it was more brilliant than in the vapour of mercury at common temperatures: the lights were of different colours; being of a pure white in the vapour of the chloride, and of a red, inclined to purple, in that of the oil; and in both cases permanent elastic fluid was produced by its transmission.

The law of the diminution of the density of vapours by diminution of temperature, has not been accurately ascertained; but I have no doubt, from the experiments of Mr. Dalton, and some I have made myself, that it is represented by a geometrical progression; the decrements of temperature being in arithmetical progression; and in three pure fluids that I operated upon,* the ratio seemed nearly uniform for the same number of degrees below the boiling point; and (taking intervals of 20 degrees of temperature) ·369416. Upon this datum, and considering the boiling point of mercury to be 600°, that of oil 540°, that of chloride of antimony 340°, and that of tin 5000°, all above 52°, and the elastic force of vapour of water at this temperature to be equal to raise by its pressure about ·45 parts of an inch of mercury; the relative strengths of vapour will be, for mercury 00015615, for oil 0016819, for chloride of antimony 01692, and for tin 37015, preceded by 48 zeros.†

It is not known whether the vapour from solids follows a similar law of progression as that from fluids, and these numbers are only given to show how minute the quantity of matter must be in vapours where its effects are distinct upon electrical phenomena; and how much more minute it must be in the case of mercury artificially cooled; and almost beyond imagination so in vapours from substances requiring very elevated temperatures for their ebullition.

I made some comparative experiments to ascertain whether below the freezing point of water, the diminution of the temperature of the torricellian vacuum diminished its powers of transmitting electricity, or of

* Water, chloride of phosphorus, and carburet of sulphur.

† I am obliged to Charles Babbage, Esq., F.R.S., for these calculations.

being rendered luminous by it. To about 20° this appeared to be the case; but between 20° above and 20° degrees below zero, the lowest temperature I could produce by pounded ice and muriate of lime, it seemed stationary; and as well as I could determine, the electrical phenomena were nearly of the same intensity as those produced in the vacuum above tin.

Unless the electrical machine was very active, no light was visible during the transmission of the electricity; but that this transmission took place, was evident from the luminous appearance of the rarefied air in the other parts of the syphon, and from the diminution of the repulsion of the ball of the quadrant electrometer attached to the prime conductor. When the machine was in great activity, there was a pale phosphorescent light above, and a spark on the mercury below, and brilliant light in the common vacuum. A Leyden jar *weakly* charged could not be made to transmit its electricity by explosion through the cooled torricellian vacuum, but this electricity was slowly dissipated through it; and when *strongly* charged, the spark passed through nearly as much space as in common air, and with a light visible in the shade. At all temperatures below 200°, the mercurial vacuum was a much worse conductor than highly rarefied air; and when the tube containing it was included in the exhausted receiver, its temperature being about 50°, the spark passed through a distance six times greater in the Boylean than in the mercurial vacuum.

It is evident from these general results that the light (and probably the heat) generated in electrical discharges depends *principally* on some properties or substances belonging to the ponderable matter through which it passes; but they prove likewise that space, where there is no appreciable quantity of this matter, is capable of exhibiting electrical phenomena: and under this point of view, they are favourable to the idea of the phenomena of electricity being produced by a highly subtil fluid or fluids, of which the particles are repulsive with respect to each other, and attractive of the particles of other matter. On such an abstruse question, however, there can be no demonstrative evidence. It may be assumed, as in the hypothesis of Hooke, Huygens, and Euler, that an etherial matter, susceptible of electrical affections, fills all space; or that the positive and electrical negative states may increase the force of vapour from the substances in which they exist; and there is a fact in favour of this last idea which I have often witnessed—when the voltaic discharge is made in the Boylean vacuum, either from platinum or charcoal in contact with mercury, the discharging surfaces require to be brought very near in the first instance; but the electricity may be made to pass to considerable distances through the vapour generated from the mercury or charcoal by its agency;—and when two surfaces of highly fixed metal, such as platinum or iron are used, the discharge will pass only through a very small distance, and cannot be permanently kept up.

The circumstance, that the intensity of the electrical light in the mercurial vacuum diminishes as it is cooled to a certain point, when the vapour must be of almost infinitely small density, and is then stationary, seems strongly opposed to the idea, that it is owing to any *permanent* vapour emitted constantly by the mercury. The results with tin must be regarded as more equivocal; because as this substance cannot be boiled in vacuo, it may be always suspected to have emitted a small quantity of the rare air or gas to which it has been exposed; yet, supposing this circumstance, such gas must be at least as highly expanded as the vapour from cooled mercury, and can hardly be supposed capable of affording the dense light, which the passage of the electricity of the charged Leyden phial through the vacuum produces.

When the intense heat produced by electricity is considered, and the strong attractive powers of differently electrified surfaces, and the rapidity of the changes of state, it does not seem at all improbable, that the superficial particles of bodies, which, when detached by the repulsive power of heat, form vapour, may be likewise detached by electrical powers, and that they may produce luminous appearances in a vacuum, free from all other matter by the annihilation of their opposite electrical states.

In common cases of electrical action, the quantity of the heat generated by the annihilation of the different electrical states depends, as I stated in my last communication to the Society, upon the nature of the matter on which it acts; and in cases when electrical sparks are taken in fluids, vapour or gas is always generated; and in elastic fluids, the intensity of the light is always greater, the denser the medium. The luminous appearances therefore, it is evident from all the statements, must be considered as secondary; whilst the uniform exertions of electrical attractions and repulsions, under all circumstances, in rare and dense media and in vacuo, and with respect to solids, fluids, and gases, point them out (whether they be specific affections of a subtile imponderable fluid, or peculiar properties of matter) as primary and invariable electrical phenomena.

PLATE II.

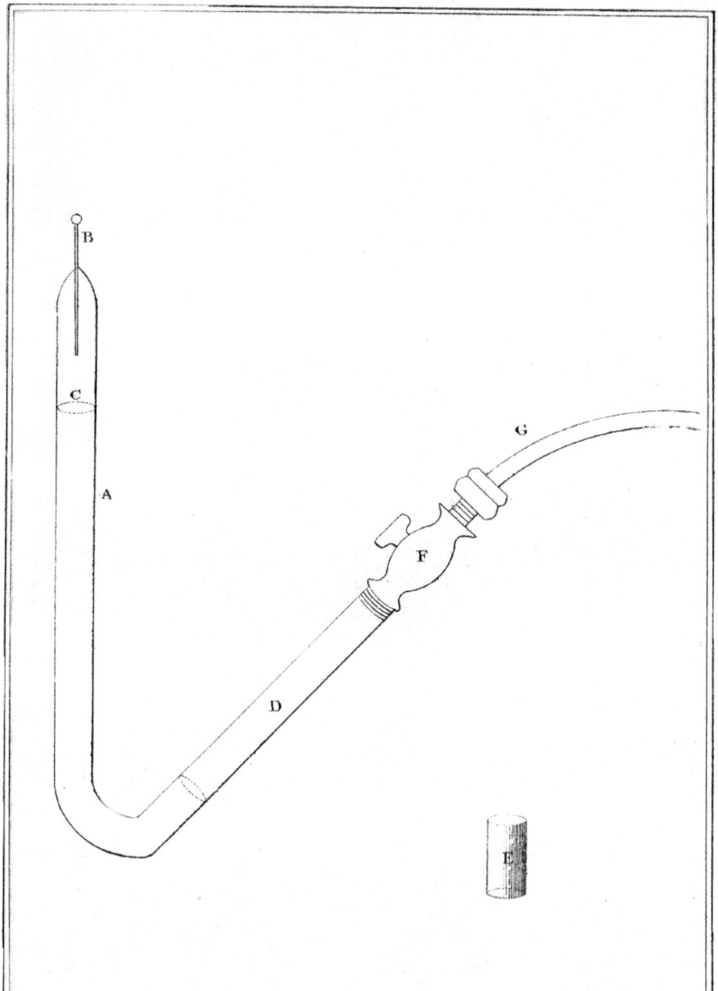

I have mentioned in the last page the suspicion that melted tin may contain air. I shall conclude this paper by stating the grounds of this suspicion, and noticing a circumstance which appears to be of considerable importance, both in relation to the construction of barometers and thermometers, and to the analysis of gaseous bodies. Recently distilled mercury that has been afterwards boiled and cooled in the atmosphere, and which presents a perfectly smooth surface in a barometer tube, emits air when strongly heated in vacuo, and that in quantities sufficient to cover the whole interior of the tube with globules; and on keeping the stop-cock of one of the tubes used in the experiments on the mercurial vacuum open for some hours, it was found that the lower stratum of mercury had imbibed air, for when heated in vacuo, it emitted it distinctly from a space of a quarter of an inch of the column; and its production ceased at about an inch high in the tube. There is great reason to believe that this air exists in mercury in the same invisible state as in water, that is, distributed through its pores; and the fact shows the necessity of long boiling the mercury in barometer and thermometer tubes, and the propriety of exposing as small a surface of mercury as possible to the air. It may explain, likewise, the difference of the heights of the mercury in different barometers, and seems to indicate the propriety of re-boiling the mercury in these instruments after a certain lapse of time.

EXPLANATION OF THE PLATE.

A. The tube of the usual diameter.
B. The wire for communicating electricity.
E. A small cylinder of metallic foil to place as a cap

on tubes not having the wire *B*, to make a coated surface.

C. The surface of the quicksilver or fused tin.

D. The part of the tube to be exhausted by the stop-cock *F*, after being filled by means of the same stop-cock, when necessary, with hydrogen.

G. The moveable tube connected with the air-pump.

It is evident that by introducing more mercury, the leg *D* may be filled with mercury, and the stop-cock closed upon it, so as to leave only a torricellian vacuum in the tube, in which the mercury may be boiled. I have found that the experiments tried in this way, offer no difference of result.

XIII.

ON A NEW PHENOMENON OF ELECTRO-MAGNETISM.*

On a subject so obscure as electro-magnetism, and connected by analogies more or less distinct with the doctrines of heat, light, electricity, and chemical attraction, it is not difficult to frame *hypotheses;* but the science is in a state too near its infancy to expect the development of any satisfactory *theory;* and its progress can only be ensured by new facts and experiments, which may prepare the way for extensive and general reasonings upon its principles.† Influenced by this opinion, I am induced to lay before the Society an account of an electro-magnetic phenomenon I observed about fifteen months ago in the laboratory of the Royal Institution, and which I have lately had an occasion of witnessing in

* [From the Phil. Trans. for 1823. Read before the Royal Society, March 6, 1823.]

† [The above distinction between *hypothesis* and *theory* was one of the strongest features of the author's philosophical method,—using the former term in the sense of supposition or opinion,— the latter in that of a generalization of facts; and considering the former only useful so far as leading to the latter: that volcanic action depends on the combustion of the metals of the earths and alkalies, beneath the surface, he held to be an hypothesis, because the existence of these metals so situated has not yet been proved to be matter of fact; that fluorine exists, he held also to be hypothetical, on the same ground; but that chlorine is a simple substance, muriatic acid gas a compound of this substance and hydrogen, the chlorides compounds of the same substance and the metals, he held to be theory, i. e., an expression of facts or the results of experiments.]

a more perfect manner, through the kindness of Mr. Pepys, by the use of a large battery, constructed under his directions for the London Institution, and containing a pair of plates of about two hundred square feet. In describing this phenomenon, I shall not enter into very minute details, because the experiments, which led to the discovery of it, are very simple, and, though more distinct with a large apparatus, yet it may be observed by the use of a pair of plates containing from ten to fifteen square feet.

Immediately after Mr. Faraday had published his ingenious experiments on electro-magnetic rotation, I was induced to try the action of a magnet on mercury connected in the electrical circuit, hoping that, in this case, as there was no mechanical suspension of the conductor, the appearances would be exhibited in their most simple form; and I found that when two wires were placed in a basin of mercury perpendicular to the surface, and in the voltaic circuit of a battery with large plates, and the pole of a powerful magnet held either above or below the wires, the mercury immediately began to revolve round the wire as an axis, according to the common circumstances of electro-magnetic rotation, and with a velocity exceedingly increased when the *opposite* poles of two magnets were used, one above, the other below.

Masses of mercury of several inches in diameter were set in motion, and made to revolve in this manner, whenever the pole of the magnet was held near the perpendicular of the wire; but when the pole was held above the mercury between the two wires, the circular motion ceased; and currents took place in the mercury in opposite directions, one to the right, and the other to the left of the magnet. These circumstances, and various others which it would be tedious to detail, induced me to believe that the passage of the electricity through the mercury produced motions independent of the action of the magnet; and that the appearances which I have described were owing to a composition of forces.

I endeavoured to ascertain the existence of these motions in the mercury, by covering its surface with weak acids; and diffusing over it finely divided matter, such as the seeds of lycopodium, white oxide of mercury, &c. but without any distinct result. It then occurred to me, that from the position of the wires, currents, if they existed, must occur chiefly in the lower, and not the upper surface of the mercury: and I consequently inverted the form of the experiment. I had two copper wires, of about one-sixth of an inch in diameter, the extremities of which were flat and carefully polished, passed through two holes three inches apart in the bottom of a glass basin, and perpendicular to it; they were cemented into the basin, and made non-conductors by sealing-wax, except at their polished ends; the basin was then filled with mercury, which stood about a tenth or twelfth of an inch above the wires. The wires were now placed in a powerful voltaic circuit. The moment the contacts were made, *the phenomenon*, which is the principal object of this paper, occurred: the mercury was immediately seen in violent agitation; its surface became elevated into a small cone above each of the wires: waves flowed off in all directions from these cones; and the only point of rest was apparently where they met in the centre of the mercury between the two wires. On holding the pole of a powerful bar magnet at a considerable distance (some inches) above one of the cones, its apex was diminished and its base extended : by lowering the pole further, those effects were still further increased, and the undulations were feebler. At a smaller distance the surface of the mercury became plane; and rotation slowly began round the wire. As the magnet approached, the rotation became more rapid, and when it was about half an inch above the mercury, a great depression of it was observed above the wire, and a vortex, which reached almost to the surface of the wire.

In the first experiments which I made, the conical elevations or fountains of mercury were about the tenth or twelfth of an inch high, and the vortices apparently as low; but in the experiments made at the London Institution, the mercury being much higher above the wire, the elevations and depressions were much more considerable, amounting to the fifth or sixth of an inch. Of course, the rotation took place with either pole of a magnet or either wire, or both together, according to the well known circumstances which determine these effects.

To ascertain whether the communication of heat diminishing the specific gravity of the mercury, had any share in these phenomena, I placed a delicate thermometer above one of the wires in the mercury, but there was no immediate elevation of temperature; the heat of the mercury gradually increased, as did that of the wires; but this increase was similar in every part of the circuit. I proved the same thing more distinctly, by making the whole apparatus a *thermometer* terminating in a fine tube filled with mercury. At the first instant that the mercury became electro-magnetic, there was no increase of its volume.

This phenomenon cannot be attributed to common electrical repulsion; for in the electro-magnetic circuit, similar electrified conductors do not repel, but attract each other; and it is in the case in which conductors in *opposite* states are brought near each other on surfaces of mercury, that repulsion takes place.

Nor can the effect be referred to that kind of action which occurs when electricity passes from good into bad conductors, as in the phenomena of points electrified in air, as the following facts seem to prove. Steel wires were substituted for copper wires, and the appearances were the same in kind, and only less in degree; without doubt, in consequence of a smaller quantity of electricity passing through the steel wires: and by comparing the conducting powers of equal cylinders of mercury and steel in glass tubes by ascertaining the quantity of iron filings they attracted, it was found that the conducting powers of mercury were higher than those of steel; the first metal taking up fifty-eight grains of iron filings, and the second only thirty-seven.

Again; fused tin was substituted for mercury in a porcelain vessel into which wires of copper and steel were alternately ground and fixed; the elevations were produced as in the mercury, and the phenomena of rotation by the magnet; and it was found by direct experiment, that the conducting powers of the tin, at and just before its point of fusion, did not perceptibly differ, and that they were much higher than those of mercury. Lastly, the communication was made from the battery by two tubes having nearly the same diameter as the wires, filled with mercury, so that the electricity, for some inches before it entered the basin, passed through mercury; and still the appearances continued the same.

From the rapidity of the undulations round the points of the cones, I thought they would put in motion any light bodies placed above mercury; but I could not produce the slightest motion in a very light wheel hung on an axle; and when fine powders of any kind were strewed upon the surface, they merely underwent undulations, without any other change of place; and fine iron filings strewed on the top of the cone, arranged themselves in right lines at right angles to the line joining the two wires, and remained stationary, even on the centre of the cone. The effect, therefore, is of a novel kind, and in one respect seems analagous to that of the tides. It would appear as if the passage of the electricity diminished the action of gravity on the mercury. And that there is no change of volume of the whole mass of the mercury appears from the experiment, page 260; and this was shown likewise by enclosing the apparatus in a kind of manometer, terminating in a fine tube containing air enclosed by oil; and which, by its expansion or contraction, would have shown the slightest change of volume in the mercury: none however took place when the contacts were alternately made and broken, unless the circuit was uninterrupted for a sufficient time to communicate sensible heat to the mercury.

This phenomenon, in which the same effects are produced at the two opposite poles, seems strongly opposed to the idea of the electro-magnetic results being produced by the transition currents or motions of a single imponderable fluid.

On the conjectural part of the subject I shall not however enter, for the reasons stated in the beginning of this paper; but I cannot with propriety conclude, without mentioning a circumstance in the history of the progress of electro-magnetism, which, though well known to many Fellows of this Society, has, I believe, never been made public, namely, that we owe to the sagacity of Dr. Wollaston, the first idea of the possibility of the rotations of the electro-magnetic wire round its axis, by the approach of a magnet; and I witnessed, early in 1821, an unsuccessful experiment which he made to produce the effect in the laboratory of the Royal Institution.

XIV.

NOTE ON THE CONDENSATION OF MURIATIC ACID GAS INTO THE LIQUID FORM.*

In desiring Mr. Faraday to expose the hydrate of chlorine to heat in a closed glass tube, it occurred to me that one of three things would happen;—that it would become fluid as a hydrate; or that a decomposition of water would occur, and euchlorine and muriatic acid be formed; or that the chlorine would separate in a condensed state. This last result having been obtained, it evidently led to other researches of the same kind. I shall hope, on a future occasion, to detail some general views on the subject of these researches. I shall now merely mention, that by sealing muriate of ammonia and sulphuric acid in a strong glass tube, and causing them to act upon each other, I have procured liquid muriatic acid; and by substituting carbonate for muriate of ammonia, I have no doubt that liquid carbonic acid may be obtained, though in the only trial I have made the tube burst. I have requested Mr. Faraday to pursue these experiments, and to extend them to all the gases which are of considerable density, or to any extent soluble in water; and I hope soon to be able to

* [This paper, which was published in the Philosophical Transactions for 1823, followed one by Mr. Faraday, "On Fluid Chlorine," in the introduction to which he states that the author suggested to him the experiment of exposing to heat the hydrate of chlorine, with the expectation of obtaining interesting results.]

lay an account of his results, with some applications of them that I propose to make, before the Society.

I cannot conclude this note without observing, that the generation of elastic substances in close vessels, either with or without heat, offers much more powerful means of approximating their molecules than those dependent upon the application of cold, whether natural or artificial; for, as gases diminish only about $\frac{1}{420}$ in volume for every degree of Fahrenheit's scale, beginning at ordinary temperatures, a very slight condensation only can be produced by the most powerful freezing mixtures, not half as much as would result from the application of a strong flame to one part of a glass tube, the other part being of ordinary temperature: and when attempts are made to condense gases into fluids by sudden mechanical compression, the heat, instantly generated, presents a formidable obstacle to the success of the experiment; whereas, in the compression resulting from their slow generation in close vessels, if the process be conducted with common precautions, there is no source of difficulty or danger; and it may be easily assisted by artificial cold in cases when gases approach near to that point of compression and temperature at which they become vapours.

XV.

ON THE APPLICATION OF LIQUIDS FORMED BY THE CONDENSATION OF GASES AS MECHANICAL AGENTS.*

ONE of the principal objects that I had in view, in causing experiments to be made on the condensation of different gaseous bodies, by generating them under pressure, was the hope of obtaining vapours, which, from the facility with which their elastic forces might be diminished or increased, by small decrements or increments of temperature, would be applicable to the same purposes as steam.

As soon as I had obtained muriatic acid in the liquid state, a body which M. Berthollet supposed owed its power of being separated from bases by other acids, only to the facility with which it assumes the gaseous form, I had no doubt, as I mentioned in my last communication, that all the other gases which have weaker affinities or

* [From the Phil. Trans. for 1823. Read before the Royal Society, April 17, 1823.]

[This paper also followed one by Mr. Faraday, "On the Condensation of several Gases into Liquids:" in commencing it, Mr. Faraday remarks, "I had the honour, a few weeks since, of submitting to the Royal Society a paper on the reduction of chlorine to the liquid state. An important note was added to the paper by the President, on the general application of the means used in this case to the reduction of other gaseous bodies to the liquid state; and in illustration of the process, the production of liquid muriatic acid was described. Sir Humphry Davy did me the honour to request I would continue the experiments, which I have done under his general direction."]

greater densities, and which are absorbable to any extent by water, might be rendered fluid by similar means; and, that the conjecture was founded, has been proved by experiments made with so much industry and ingenuity, and which I have had the pleasure of communicating to the Society.

The elasticity of vapours in contact with the liquids from which they are produced, under high pressures, by high temperatures, such as those of alcohol and water, is known to increase in a much higher ratio than the arithmetical one of the temperature; but the exact law is not yet determined; and the result is a complicated one, and depends upon circumstances which require to be ascertained by experiment. Thus the ratio of the elastic force, dependent upon pressure, is to be combined with that of the expansive force dependent upon temperature; and the greater loss of radiant heat at high temperatures, and the development of latent heat in compression, and the necessity for its re-absorption in expansion (as the rationale of the subject is at present understood) must awaken some doubts as to the economical results to be obtained by employing the steam of water under very great pressures, and at very elevated temperatures.

No such doubt, however, can arise with respect to the use of such liquids, as require for their existence even a compression equal to that of the weight of thirty or forty atmospheres; and where common temperatures, or slight elevations of them, are sufficient to produce an immense elastic force; and when the principal question to be discussed, is, whether the effect of mechanical motion is to be most easily produced by an increase or diminution of heat by artificial means.

With the assistance of Mr. Faraday I have made some experiments on this subject, and the results have answered my most sanguine expectations. Sulphuretted hydrogen, which condenses readily at 3° F., under a pressure equal to that which balances the elastic force of an atmosphere compressed to $\frac{1}{14}$, had its elastic force increased so as to equal that of an atmosphere compressed to $\frac{1}{17}$ by an increase of 47° of temperature. Liquid muriatic acid at 3°, exerted an elastic force equivalent to that of an atmosphere compressed to $\frac{1}{20}$; by an increase of 22°, it gained an elastic force equivalent to that of an atmosphere compressed to $\frac{1}{25}$; and by a farther addition of 26°, an elastic force equivalent to that of air condensed to $\frac{1}{40}$ of its primitive volume. These experiments were made in thick glass tubes hermetically sealed. The degree of pressure was estimated by the change of volume of air confined by mercury in a small graduated gage, and placed in a part of the tube exposed to the atmosphere, and the temperatures were diminished from the degree at which the gage was introduced, *i. e.* the atmospheric temperature, by freezing mixtures; so that the temperature of the air within the gage could not be considerably altered; and as the elastic fluid surrounding the gage must have had a higher temperature than the condensed fluid, the diminution of the elastic force of the vapour from the fluids cannot be considered as over-rated.

From the immense differences between the increase of elastic force in gases under high and low pressures by similar increments of temperature, there can be no doubt that the denser the vapour, or the more difficult of condensation the gas, the greater will be its power under changes of temperature as a mechanical agent: thus, carbonic acid will be much more powerful than muriatic acid. In the only experiment which has been tried upon it, its force was found to be nearly equal to that of air compressed to $\frac{1}{20}$ at 12° F., and of air compressed to $\frac{1}{36}$ at 32°, making an increase equal to the weight of 13 atmospheres by an increase of 20° of temperature; and this immense elastic force of 36 atmospheres being exerted at the freezing point of water.* And azote, if it could be obtained fluid, would, there is no doubt, be far more powerful than carbonic acid; and hydrogen, in such a state, would exert a force almost incalculably great, and liable to immense changes from the slightest variations of temperature.

To illustrate this idea, I shall quote an experiment on alcohol of sulphur.

The temperature of this body was raised 20° above its boiling point, and its elastic force examined; it was found equal to less than that of air compressed to $\frac{3}{4}$. It was now heated to 320° under a pressure equal to that of air condensed to $\frac{10}{77}$, and a similar increment of 20° added; its elastic force became equivalent to that of an atmosphere compressed to $\frac{100}{895}$.

I hope soon to be able to repeat these experiments,† in a more minute and accurate way; but the general results appear so worthy the attention of practical mechanics, that I think it a duty to lose no time in bring-

* Since this paper was read, Mr. Faraday has ascertained that the vapour of ammonia at 32° exerts an elastic force equal to that of an atmosphere compressed to $\frac{1}{5}$; and at 50° to that of an atmosphere compressed to $\frac{10}{65}$; and that the vapour of nitrous oxide, at 32°, has an elastic force equal to that of an atmosphere compressed to $\frac{1}{44}$; and at 45° to an atmosphere compressed to $1{\cdot}0 \div 51{\cdot}3$ nearly.

† [This intention, the author, I believe, did not carry into effect, nor has it been attempted that I am aware of by others, notwithstanding the incalculable importance of the object in view—the obtaining a fluid applicable to the steam engine, capable of producing the effects of ordinary steam, without the consumption of fuel; he was probably diverted from the inquiry by another very important one, which about this time he engaged in, namely, the protection of the copper sheathing of ships.]

ing them forward even in their present imperfect state.

In applying the condensed gases as mechanical agents, there will be some difficulty; the materials of the apparatus must be at least as strong and as perfectly joined as those used by Mr. Perkins in his high pressure steam engine: but the small differences of temperature required to produce an elastic force equal to the pressure of many atmospheres, will render the risk of explosion extremely small; and if future experiments should realise the views I have developed, the mere difference of temperature between sunshine and shade, and air and water, or the effects of evaporation from a moist surface, will be sufficient to produce results, which have hitherto been obtained only by a great expenditure of fuel.

I shall conclude this communication by a few general observations arising out of this inquiry.

There is a simple mode of liquefying the gases, which at first view appears paradoxical; namely, by the application of *heat:* it consists in placing them in one leg of a bent sealed tube confined by mercury, and applying heat to ether, or alcohol, or water, in the other end. In this manner, by the pressure of the vapour of ether, I have liquefied prussic gas and sulphurous acid gas, the only two on which I have made experiments; and these gases in being reproduced occasioned cold.

There can be little doubt that these general facts of the condensation of the gases will have many practical applications. They offer easy methods of impregnating liquids with carbonic acid and other gases, without the necessity of common mechanical pressure.

They afford means of producing great diminutions of temperature, by the rapidity with which large quantities of liquids may be rendered aëriform; and as compression occasions similar effects to cold, in preventing the formation of elastic substances, there is great reason to believe that it may be successfully employed for the preservation of animal and vegetable substances for the purposes of food.

APPENDIX TO THE PRECEDING PAPER.

On the Changes of Volume produced in Gases in different States of Density by Heat.

In investigating the laws of the elastic forces exerted by vapours or gases raised from liquids by increase or temperature under compression, one of the most important circumstances to be considered is the rate or expansion, or what is equivalent, of the elastic force, in atmospheres in different states of density.

It has been shown by the experiments of Mr. Dalton and of M. Gay Lussac, that elastic fluids of very different specific gravities expand equally by equal increments of temperature, or, as may be more correctly expressed, according to the elucidations of MM. Dulong and Petit, that mercury and air, or gases, are equivalent in their expansions for any number of degrees in the thermometrical scale between the freezing and boiling points of water; and the early researches of M. Amontons seemed to show that the increase of the spring or elastic force of air by increase of temperature, was in the direct ratio of its density. I am not, however, acquainted with any direct researches upon the changes of volume produced in gases in very different states of condensation and rarefaction by changes of temperature; and the importance of the inquiry, in relation to the subject of my last communication to the Society, induced me to undertake the following experiments.

Dry atmospherical air was included in a tube by mercury, and its temperature raised from 32° F., to 212°, and its expansion accurately marked. The same volumes of air, but of double and of more than triple the density under a pressure of thirty and sixty-five inches of mercury, were treated in the same manner, and in the same tubes; and when the necessary corrections were made for the difference of pressure of the removed column of mercury, it was found that the expansions were exactly the same.

An apparatus was constructed in which the expansions of rare air, confined by columns of mercury, were examined and compared with the expansions of equal volumes of air under common pressure; when it appeared, that for an equal number of degrees of Fahrenheit's scale, and between 32° and 212°, they were precisely equal, whether the air was one-half, one-third, or one-sixth of its natural density.

Similar experiments were made, but they were necessarily less precise, with air condensed six and expanded fifteen times, with similar results.

XVI.

ON THE CORROSION OF COPPER SHEATHING BY SEA-WATER, AND ON METHODS OF PREVENTING THIS EFFECT; AND ON THEIR APPLICATION TO SHIPS OF WAR AND OTHER SHIPS.*

I. THE rapid decay of the copper sheathing of His Majesty's ships of war, and the uncertainty of the time of its duration, have long attracted the attention of those persons most concerned in the naval interests of the country. Having had my inquiries directed to this important object by the Commissioners of the Navy Board, and a Committee of the Royal Society having been appointed to consider of it, I entered into an experimental investigation of the causes of the action of sea-water upon copper. In pursuing this investigation, I have ascertained many facts which I think not unworthy of the notice of the Royal Society, as they promise to illustrate some obscure parts of electro-chemical science; and likewise seem to offer important practical applications.

II. It has been generally supposed that sea-water had little or no action on pure copper, and that the rapid decay of the copper on certain ships was owing to its impurity. On trying, however, the action of sea-water upon two specimens of copper, sent by John Vivian, Esq. to Mr. Faraday for analysis, I found the specimen which appeared absolutely pure, was acted upon even

[* From the Philosophical Transactions for 1824. Read before the Royal Society, January 22, 1824.]

more rapidly than the specimen which contained alloy: and, on pursuing the inquiry with specimens of various kinds of copper which had been collected by the Navy Board, and sent to the Royal Society, and some of which had been considered as remarkable for their durability, and others for their rapid decay, I found that they offered very inconsiderable differences only in their action upon sea-water; and, consequently, that the changes they had undergone must have depended upon other causes than the absolute quality of the metal.

III. To enable persons to understand fully the train of these researches, it will be necessary for me to describe the nature of the chemical changes taking place in the constituents of sea-water by the agency of copper.

When a piece of polished copper is suffered to remain in sea-water, the first effects observed are, a yellow tarnish upon the copper, and a cloudiness in the water, which take place in two or three hours: the hue of the cloudiness is at first white; it gradually becomes green. In less than a day a bluish-green precipitate appears in the bottom of the vessel, which constantly accumulates; at the same time that the surface of the copper corrodes, appearing red in the water, and grass-green where it is in contact with air. Gradually carbonate of soda forms upon this grass-green matter; and these changes continue till the water becomes much less saline.

The green precipitate, when examined by the action of solution of ammonia and other tests, appears principally to consist of an insoluble compound of copper, (which may be considered as a hydrated sub-muriate) and hydrate of magnesia.

According to the views which I developed fourteen years ago, of the nature of the compounds of chlorine, and which are now generally adopted, it is evident that soda and magnesia cannot appear in sea-water by the action of a metal, unless in consequence of an absorption or transfer of oxygen. It was therefore necessary for these changes, either that water should be decomposed, or oxygen absorbed from the atmosphere. I found that no hydrogen was disengaged, and consequently no water decomposed: necessarily, the oxygen of the air must have been the agent concerned, which was made evident by many experiments.

Copper in sea-water deprived of air by boiling or exhaustion, and exposed in an exhausted receiver or an atmosphere of hydrogen gas, underwent no change; and an absorption in atmospherical air was shown when copper and sea-water were exposed to its agency in close vessels.

IV. In the Bakerian Lecture for 1806, I have advanced the hypothesis, that chemical and electrical changes may be identical, or dependent upon the same property of matter: and I have farther explained and illustrated this hypothesis in an elementary work on chemistry, published in 1812. Upon this view, which has been adopted by M. Berzelius and some other philosophers, I have shown that chemical attractions may be exalted, modified, or destroyed, by changes in the electrical states of bodies; that substances will only combine when they are in different electrical states; and that, by bringing a body naturally positive artificially into a negative state, its usual powers of combination are altogether destroyed; and it was by an application of this principle that, in 1807, I separated the bases of the alkalies from the oxygen with which they are combined, and preserved them for examination; and decomposed other bodies formerly supposed to be simple.

It was in reasoning upon this general hypothesis likewise, that I was led to the discovery which is the subject of this paper.

Copper is a metal only weakly positive in the electrochemical scale; and, according to my ideas, it could only act upon sea-water when in a positive state; and, consequently, if it could be rendered slightly negative, the corroding action of sea-water upon it would be null; and whatever might be the differences of the kinds of copper sheathing and their electrical action upon each other, still every effect of chemical action must be prevented, if the whole surface were rendered negative. But how was this to be effected? I at first thought of using a Voltaic battery; but this could be hardly applicable in practice. I next thought of the contact of zinc, tin, or iron: but I was for some time prevented from trying this, by the recollection that the copper in the Voltaic battery, as well as the zinc, is dissolved by the action of diluted nitric acid; and by the fear that too large a mass of oxidable metal would be required to produce decisive results. After reflecting, however, for some time on the slow and weak action of sea-water on copper, and the small difference which must exist between their electrical powers; and knowing that a very feeble chemical action would be destroyed by a very feeble electrical force, I resolved to try some experiments on the subject. I began with an extreme case. I rendered sea-water slightly acidulous by sulphuric acid, and plunged into it a polished piece of copper, to which a piece of tin was soldered equal to about one-twentieth of the surface of the copper. Examined after three days the copper remained perfectly clean, whilst the tin was rapidly corroded: no blueness appeared in this liquor; though, in a comparative experiment, when *copper alone* and the same fluid mixture was used, there was a considerable corrosion of the copper, and a distinct blue tint in the liquid.

If one-twentieth part of the surface of tin prevented the action of sea-water rendered slightly acidulous by sulphuric acid, I had no doubt that a much smaller quantity would render the action of sea-water, which depended only upon the loosely attached oxygen of common air, perfectly null; and on trying $\frac{1}{200}$ part of tin, I found *the effect* of its preventing the corrosion of the copper perfectly decisive.

V. This general result being obtained, I immediately instituted a number of experiments, in most of which I was assisted by Mr. Faraday, to ascertain all the circumstances connected with the preservation of copper by a more oxidable metal. I found, that whether the tin was placed either in the middle, or at the top, or at the bottom of the sheet of copper, its effects were the same; but, after a week or ten days, it was found that the defensive action of the tin was injured, a coating of sub-muriate having formed, which preserved the tin from the action of the liquid.

With zinc or iron, whether malleable or cast, no such diminution of effect was produced. The zinc occasioned only a white cloud in the sea-water, which speedily sunk to the bottom of the vessel in which the experiment was made. The iron occasioned a deep orange precipitate: but after many weeks, not the smallest portion of copper was found in the water; and so far from its surface being corroded, in many parts there was a regeneration of zinc or of iron found upon it.

VI. In pursuing these researches, and applying them to every possible form and connection of sheet copper, the results were of the most satisfactory kind. A piece of zinc as large as a pea, or the point of a small iron nail, were found fully adequate to preserve forty or fifty square inches of copper; and this, wherever it was placed, whether at the top, bottom, or in the middle of the sheet of copper, and whether the copper was straight or bent, or made into coils. And where the connection between different pieces of copper was completed by wires, or thin filaments of the fortieth or fiftieth of an inch in diameter, the effect was the same; every side, every surface, every particle of the copper remained bright, whilst the iron or the zinc was slowly corroded.

A piece of thick sheet copper, containing on both sides about sixty square inches, was cut in such a manner as to form seven divisions, connected only by the smallest filaments that could be left, and a mass of zinc, of the fifth of an inch in diameter, was soldered to the upper division. The whole was plunged under sea-water; the copper remained perfectly polished. The same experiment was made with iron: and now, after a lapse of a month, in both instances, the copper is as bright as when it was first introduced, whilst similar pieces of copper, undefended, in the same sea water, have undergone considerable corrosion, and produced a large quantity of green deposit in the bottom of the vessel.

A piece of iron nail about an inch long was fastened by a piece of copper wire, nearly a foot long, to a mass of sheet copper, containing about forty square inches, and the whole plunged below the surface of sea-water: it was found, after a week, that the copper was defended by the iron in the same manner as if it had been in immediate contact.

A piece of copper and a piece of zinc soldered together at one of their extremities, were made to form an arc in two different vessels of sea-water; and the two portions of water were connected together by a small mass of tow moistened in the same water; the effect of the preservation of the copper took place in the same manner as if they had been in the same vessel.

As the ocean may be considered, in its relation to the quantity of copper in a ship, as an infinitely extended conductor, I endeavoured to ascertain whether this circumstance would influence the results, by placing two very fine copper wires, one undefended, the other defended by a particle of zinc, in a very large vessel of sea-water, which water might be considered to bear the same relation to so minute a portion of metal as the sea to the metallic sheathing of a ship. The result of this experiment was the same as that of all the others; the defended copper underwent no change; the undefended tarnished, and deposited a green powder.

Small pieces of zinc were soldered to different parts of a large plate of copper, and the whole plunged in sea-water: it was found that the copper was preserved in the same manner as if a single piece had been used.

A small piece of zinc was fastened to the top of a plate of polished copper, and a piece of iron of a much larger size was soldered to the bottom, and the combination placed in sea-water. Not only was the copper preserved on both sides in the same manner as in the other experiments, but even the iron; and after a fortnight, both the polish of copper and the iron remained unimpaired.

VII. I am continuing these researches, and I shall communicate such of them as are connected with new facts, to the Royal Society.

The Lords Commissioners of the Admiralty, with their usual zeal for promoting the interests of the Navy by the application of science, have given me permission to ascertain the practical value of these results, by experiments upon ships of war; and there seems every reason to expect (unless causes should interfere of which our present knowledge gives no indications) that small quantities of zinc, or, which is much cheaper, of malleable, or cast iron, placed in contact with the copper sheathing of ships, which is all in electrical connection, will entirely prevent its corrosion. And as negative electricity cannot be supposed favourable to animal or vegetable life; and as it occasions the deposition of magnesia, a substance exceedingly noxious to land vegetables, upon the copper surface: and as it must assist in preserving its polish, there is considerable ground for hoping that the same application will keep the bottoms of ships clean, a circumstance of great importance both in trade and naval war.

It will be unnecessary for me to dwell upon the economical results of this discovery, should it be successful in actual practice, or to point out its uses in this great maritime and commercial country.

I might describe other applications of the principle to the preservation of iron, steel, tin, brass, and various useful metals; but I shall reserve this part of the subject for another communication to the Royal Society.

XVII.

ADDITIONAL EXPERIMENTS AND OBSERVATIONS ON THE APPLICATION OF ELECTRICAL COMBINATIONS TO THE PRESERVATION OF THE COPPER SHEATHING OF SHIPS, AND TO OTHER PURPOSES.*

I HAVE already had the honour of communicating to the Royal Society the results of my first researches on the modes of preventing the chemical action of fluid menstrua, such as saline solutions, or sea-water, containing air on copper, by the contact of more oxidable metals.

For some months I have been engaged in a series of experiments on this subject, so important to the navigation and commerce of the country: and through the liberal and enlightened views of Lord Melville and the Lords of the Admiralty, who desired the Commissioners of the Navy Board and of the Dock Yards to give me every assistance in their power, and all the facilities which our magnificent Naval Establishments at Chatham and Portsmouth furnish, I have been enabled to conduct my operations upon a very large scale. At this advanced period of the Session, it will be impossible for me to give more than a very short notice of experiments which have been tried under a great variety of circumstances, and the details of which would occupy some hours in reading; but I cannot deprive myself of the

* [From the Phil. Trans. for 1824. Read before the Royal Society, June 17, 1824.]

pleasure of stating the satisfactory and conclusive nature of the results, many of which have even surpassed my expectations.

Sheets of copper, defended by from $\frac{1}{40}$ to $\frac{1}{1000}$ part of their surface of zinc, malleable and cast-iron, have been exposed, for many weeks, in the flow of the tide in Portsmouth Harbour, and their weights ascertained before and after the experiment. When the metallic protector was from $\frac{1}{40}$ to $\frac{1}{150}$, there was no corrosion nor decay of the copper; with smaller quantities, such as from $\frac{1}{200}$ to $\frac{1}{400}$, the copper underwent a loss of weight, which was greater in proportion as the protector was smaller; and as a proof of the universality of the principle, it was found that even $\frac{1}{1000}$ part of cast-iron saved a certain proportion of the copper.

The sheathing of boats and ships, protected by the contact of zinc, cast and malleable iron in different proportions, compared with those of similar boats and sides of ships unprotected, exhibited bright surfaces—whilst the unprotected copper underwent rapid corrosion, becoming first red, then green, and losing a part of its substance in scales.

Fortunately, in the course of these experiments, it has been proved that cast-iron, the substance which is cheapest and most easily procured, is likewise most fitted for the protection of the copper. It lasts longer than malleable iron, or zinc; and the plumbaginous substance, which is left by the action of sea-water upon it, retains the original form of the iron, and does not impede the electrical action of the remaining metal.

I had anticipated the deposition of alkaline substances in certain cases, upon the negatively-electrical copper. This has actually happened. Some sheets of copper, that have been exposed nearly four months to the action of sea-water, defended by from $\frac{1}{35}$ to $\frac{1}{80}$ of their surface of zinc and iron, have become coated with a white matter, which, on analysis, has proved to be principally carbonated lime, and carbonate and hydrate of magnesia. The same thing has occurred with two harbour boats, one of which was defended by a band of zinc, the other by a band of iron, equal to about $\frac{1}{35}$ of the surface of the copper.

These sheets and boats remained perfectly clean for many weeks, as long as the metallic surface of the copper was exposed; but lately, since it has become coated with carbonate of lime and magnesia, weeds have adhered to these coatings, and insects collected on them; but on the sheets of copper, defended by quantities of cast-iron and zinc, bearing a proportion below $\frac{1}{150}$, the electrical power of the copper being less negative, more neutralised, and nearly in equilibrio with that of the menstruum, no such effect of deposition of alkaline matter or adherence of weeds has taken place; and the surface, though it has undergone a slight degree of solution, has remained perfectly clean: a circumstance of great importance, as it points out the *limits* of *protection;* and makes the application of a *very small* quantity of the oxidable metal, more advantageous, in fact, than that of a larger one.

The wear of cast-iron is not so rapid, but that a mass of two or three inches in thickness will last for some years. At least the consumption in experiments which have been going on for nearly four months, does not indicate a higher ratio. This must, however, depend on the relation of its mass to that of the copper, and upon other circumstances not yet ascertained, (such as temperature, the relative saltness of the sea, and perhaps the rapidity of the motion of the ship;) circumstances in relation to which I am about to make decisive experiments.

Many singular facts have occurred in the course of these researches. I shall mention some of them, that I have confirmed by repeated experiments, and which have connections with general science.

Weak solutions of salt act strongly upon copper; strong ones, as brine, do not affect it; and the reason seems to be that they contain little or no atmospheric air, the oxygen of which seems necessary to give the electro-positive principle of change to menstrua of this class.

I had anticipated the result of this experiment, and upon the same principle of some others.

Alkaline solutions, for instance, impede or prevent the action of sea-water on copper; having in themselves the positive electrical energy, which renders the copper negative. Lime-water, even in this way, renders null the power of action of copper on sea-water.*

The tendency of electrical and chemical action being always to produce an equilibrium in the electrical powers, the agency of all combinations formed of metals and fluids, is to occasion decompositions in such an order, that alkaline, metallic, and inflammable matters are determined to the negative part of the combination; and chlorine, iodine, oxygen, and acid matters to the positive part. I have shown, in the Bakerian Lecture for 1806, that this holds good in the Voltaic battery. The same law applies to these feebler combinations. If copper in contact with cast-iron be placed in a vessel half full of sea-water, and having its surface partially above that of the water, it will become coated with car-

* I am at present engaged in applying this principle to experiments on the preservation of animal and vegetable substances.

bonate of lime, carbonate of magnesia, and carbonate of soda; and the carbonate of soda will gradually accumulate, till the whole surface in the air is covered with its crystals: and if the iron is in one vessel, and the copper forming an arc with it in another; and a third vessel of sea-water in electrical connection, by asbestos or cotton, is intermediate, the water in this intermediate vessel continually becomes less saline; and undoubtedly, by a continuance of the process, might be rendered fresh.

I shall not take up the time of the Society, by referring to some obvious practical applications of these researches to the preservation of finely divided astronomical instruments of brass by iron, of instruments of steel by iron, or zinc: my friend Mr. Pepys has already ingeniously taken advantage of this last circumstance, in enclosing finely-cutting instruments in handles or cases lined with zinc, and many other such applications will occur. I cannot conclude, without mentioning particularly my obligations to Sir Byam Martin, the Comptroller, and Sir Robert Seppings, the Surveyor of the Navy, for the interest they have taken, and the zeal they have shown in promoting these researches; and without stating how much I owe to the care, attention, and accuracy of Mr. Nolloth, Master Shipwright, and Mr. Goodrich, Mechanist in the Dock-yard at Portsmouth, in superintending the execution of many of the experiments.

XVIII.

FURTHER RESEARCHES ON THE PRESERVATION OF METALS BY ELECTRO-CHEMICAL MEANS.*

In two papers read before the Royal Society, I have described the effects of small quantities of electro-positive metals in preventing the corrosion or chemical changes of copper exposed to sea-water, and I have stated that the results appear to be of the same kind, whether the experiments are made upon a minute scale, and in confined portions of water, or on large masses, and in the ocean.

The first and preliminary experiments proved, that the copper sheathing of ships might be preserved by this method; but another and a no less important circumstance was to be attended to, how far the cleanness of the bottom, or its freedom from the adhesion of weeds or shell-fish, would be influenced by this preservation.

The use of the copper sheathing on the bottom of ships is two-fold: first, to protect the wood from destruction by worms.

And secondly, to prevent the adhesion of weeds, barnacles, and other shell-fish. No worms can penetrate the wood as long as the surface of the copper remains perfect; but when copper has been applied to the bottom of a ship for a certain time, a green coating or rust, consisting of oxide, submuriate and carbonate of

* [From the Phil. Trans. for 1825. Read before the Royal Society June 9th, 1825.]

copper and carbonate of magnesia forms upon it, to which weeds and shell fish adhere.

As long as the whole surface of the copper changes or corrodes, no such adhesions can occur; but wheu this green rust has partially formed, the copper below is protected by it, and there is an unequal action produced, the electrical effect of the oxide, submuriate, and carbonate of copper formed, being to produce a more rapid corrosion of the parts still exposed to sea-water; so that the sheets are often found perforated with holes in one part, after being used five or six years, and comparatively sound in other parts.

There is nothing in the poisonous nature of the metal which prevents these adhesions. It is *the solution* by which they are *prevented — the wear* of surface. Weeds and shell-fish readily adhere to the poisonous salts of lead which form upon the lead protecting the fore part of the keel; and to the copper, in any chemical combination in which it is insoluble.

In general, in ships in the navy, the first effect of the adhesion of weeds is perceived upon the heads of the mixed metal nails, which consist of copper alloyed by a small quantity of tin. The oxides of tin and copper which form upon the head of the nail and in the space round it, defend the metal from the action of sea-water; and being negative with respect to it, a stronger corroding effect is produced in its immediate vicinity, so that the copper is often worn into deep and irregular cavities in these parts.

When copper is unequally worn, likewise in harbours or seas where the water is loaded with mud or mechanical deposits, this mud or these deposits rest in the rough parts or depressions in the copper, and in the parts where the different sheets join, and afford a soil or bed in which sea-weeds can fix their roots, and to which zoophytes and shell-fish can adhere.

As far as my experiments have gone, small quantities of other metals, such as iron, tin, zinc, or arsenic, in alloy in copper, have appeared to promote the formation of an insoluble compound on the surface; and consequently there is much reason to believe must be favourable to the adhesion of weeds and insects.

I have referred in my last paper to the circumstance of the carbonate of lime and magnesia forming upon sheets of copper, protected by a quantity of iron above $\frac{1}{120}$ parts, when these sheets were in harbour and at rest.

The various experiments that I have caused to be made at Portsmouth, show all the circumstances of this kind of action, and I have likewise elucidated them by experiments made on a smaller scale, and in limited quantities of water. It appears from these experiments, that sheets of copper at rest in sea-water, always increase in weight from the deposition of the alkaline and earthy substances, when defended by a quantity of cast-iron under $\frac{1}{150}$ of their surface, and if in a limited or confined quantity of water, when the proportion of the defending metal is under $\frac{1}{400}$. With quantities below these respectively proportional for the sea, and limited quantities of water, the copper corrodes; at first it slightly increases in weight, and then slowly loses weight. Thus a sheet of copper four feet long, fourteen inches wide, and weighing 9lb. 6oz., protected by $\frac{1}{100}$ of its surface of cast-iron, gained in ten weeks and five days, twelve drachms, and was coated over with carbonate of lime and magnesia: a sheet of copper of the same size protected by $\frac{1}{150}$, gained only one drachm in the same time, and a part of it was green from the adhering salts of copper; whilst an unprotected sheet of the same class, both as to size and weight, and exposed for the same time, and as nearly as possible under the same circumstances, had lost fourteen drachms; but experiments of this kind, though they agree when carried on under precisely similar circumstances, must of necessity be very irregular in their results, when made in different seas and situations, being influenced by the degree of saltness, and the nature of the impregnations of the water, the strength of tide and of the waves, the temperature, &c.

In examining sheets which had been defended by small quantities of iron in proportions under $\frac{1}{250}$ and above $\frac{1}{1000}$, whether they were exposed alone, or on the sides of boats, there seemed to me no adhesions of confervæ, except in cases where the oxide of iron covered the copper immediately round the protectors: and even in these instances such adhesions were extremely trifling, and might be considered rather as the vegetations caught by the rough surface of the oxide of iron, than as actually growing upon it.

Till the month of July, 1824, all the experiments had been tried in harbour, and in comparatively still water; and though it could hardly be doubted, that the same principles would prevail in cases where ships were in motion, and on the ocean; yet still it was desirable to determine this by direct experiment; and I took the opportunity of an expedition intended to ascertain some points of longitude in the north seas, and which afforded me the use of a steam boat, to make these researches. Sheets of copper, carefully weighed, and with different quantities of protecting metal, and some unprotected, were exposed upon canvass, so as to be electrically insulated upon the bow of the steam-boat, and were weighed and examined at different periods, after being exposed in the north seas to the action of the water during the most rapid motion of the vessel. Very rough weather interfered with some of these experiments, and many of the sheets were lost, and the protectors of others were washed away; but the general results were as satisfactory as if the whole series of the arrangements had been complete. It was found that undefended sheets of copper, of a foot square, lost about 6·55 grains in passing at a rate averaging that of eight miles an hour in twelve hours: but a sheet, having the same surface, defended by rather less than $\frac{1}{500}$ lost 5·5 grains; and that like sheets defended by $\frac{1}{70}$ and $\frac{1}{100}$ of malleable iron were similarly worn, and underwent nearly the same loss, that of two grains, in passing through the same space of water. These experiments (the results of which were confirmed by those of others, made during the whole of a voyage to and from Heligoland; but in which, during the return, the protectors were lost) show that motion does not affect the nature of the limits and quantity of the protecting metal; and likewise prove, that independently of the chemical, there is a mechanical wear of the copper in sailing, and which, on the most exposed part of the ship, and in the most rapid course, bears a relation to it of nearly 2 to 4·55.

I used the very delicate balance belonging to the Royal Society in these experiments: the sheets of copper weighed between 7000 and 8000 grains; and I was fully enabled to ascertain, by means of this balance, a diminution of weight upon so large a quantity, equal to $\frac{1}{100}$ of a grain. It was evident, from a very minute inspection of the sheet with the largest quantity of protecting metal, that there was not any adhesion of alkaline or earthy substances to its surface.

Having observed, in examining the results of some of the experiments on the effects of single masses of protecting metal on the sheathing of ships, that there was, in some cases in which sheets with old fastening had been used, tarnish or corrosion, which seemed to increase with the distance from the protecting metal, it became necessary to investigate this circumstance, and to ascertain the extent of the diminution of electrical action in instances of imperfect or irregular conducting surfaces.

With single sheets or wires of copper, and in small confined quantities of sea-water, there seemed to be no indications of diminution of conducting power, or of the preservative effects of zinc or iron, however divided or diffused the surface of the copper, provided there was a perfect metallic connexion through the mass. Thus, a small piece of copper, containing about 32 square inches, was perfectly protected by a quantity of zinc, which was less than $\frac{1}{4000}$ part of the whole surface; and a copper wire, of several feet in length, was prevented from tarnishing, by a piece of zinc wire, which was less than $\frac{1}{1400}$ part of its length. In these cases the protecting metal corroded with great rapidity, and, in a few hours, was entirely destroyed; but when applied in the form of wire and covered, except at its transverse surface, with cement, its protecting influence upon the same minute scale was exhibited for many days. A part of these results depends upon the absorption of the oxygen, dissolved in the water when its quantity is limited, by the oxidable metal, and of course the proportion of this metal must be much larger when the water is constantly changing; but the experiments seem to show that any diminution of protecting effect at a distance, does not depend upon the nature of the metallic, but of the imperfect or fluid conductor.

'I his indeed is shown by many other results.

A piece of zinc and a piece of copper in the same vessel of sea-water, but not in contact, were connected by different lengths of fine silver wire of different thickness. It was found that whatever lengths of wire of $\frac{1}{300}$ of an inch were used, there was no diminution of the protecting effect of the zinc; and the experiment was carried so far as to employ the whole of a quantity of extremely fine wire, amounting to upwards of forty feet in length, and of a diameter equal only to $\frac{100}{98742}$ of an inch, when the results were precisely the same as if the zinc and copper had been in immediate contact.

Pieces of charcoal, which is the worst amongst the more perfect conductors, were connected by being tied together, and made the medium of communication between zinc and copper, upon the same principles, and with the same views as those just described, and with precisely the same consequences.

In my first experiments upon the effects of increasing the length or diminishing the mass of the imperfect or fluid conducting surface in interfering with the preserving effects of metals, I used long narrow tubes; but I found them very inconvenient; and I had recourse to the more simple method of employing cotton or tow for this purpose.

Several feet of copper wire in a spiral form were connected with a small piece of zinc wire of about half an inch in length. The zinc and a portion of the copper were introduced into one glass, and the coils of copper wire were introduced into other glasses, so as to form a series of six or seven glasses, which were filled with sea-water, and made part of the same voltaic arrangement, by being connected with pieces of tow moistened in sea-water.

It was found in these experiments, that when the pieces of tow connecting the glasses were half an inch in thickness, the preserving effect of the zinc in the first glass was no where diminished, but extended apparently equally through the whole series.

When the pieces of tow were about the fifth of an inch in thickness, a diminution of the preserving effects of the zinc was perceived in the fourth glass, in which there was a slight solution of copper; in the fifth glass this result was still more distinct, and so on till in the seventh glass there was a considerable corrosion or the copper.

When the tow was only the tenth of an inch in thickness, the preserving effect of the zinc extended only to the third glass; and in each glass more remote, the effect of corrosion was more distinct, till in the seventh glass it was nearly the same as if there had been no protecting metal. All the chemical changes dependent upon negative electricity were successively and elegantly exhibited in this experiment. In the first glass containing the zinc, there was a considerable and hasty deposition of earthy and alkaline matter, and crystals of carbonate of soda adhered to the copper at the surface where it was clean and bright; but in the lower part it was coated with revived metallic zinc. In the second glass the wire was covered over with fine crystals of carbonate of limc; and the same phenomenon of the separation of carbonate of soda occurred, but in a less degree. In the third glass the wire was clean, but without depositions; and the presence of alkaline matter could only be distinguished by chemical tests. In the fourth glass the copper was bright, evidently in consequence of a slight but general corrosion, but with a scarcely sensible deposit; in the fifth, the deposit was very visible; and in the seventh the wire was covered with green rust.

These results, which showed that a very small quantity only of the imperfect or fluid conductor was sufficient to transmit the electrical power, or to complete the chain, induced me to try if copper nailed upon wood, and protected merely by zinc or iron on the under surface, or that next the wood, would not be defended from corrosion. For this purpose I covered a piece of wood with small sheets of copper, a nail of zinc of about the $\frac{1}{200}$ part of the surface of the copper being previously driven into the wood: the apparatus was plunged in a large jar of sea-water: it remained perfectly bright for many weeks, and when examined, it was found that the zinc had only suffered partial corrosion; that the wood was moist, and that on the interior of the copper there was a considerable portion of revived zinc, so that the negative electricity, by its operation, provided materials for its future and constant excitement. In several trials of the same kind, iron was used with the same results; and in all these experiments there appeared to be this peculiarity in the appearance of the copper, that unless the protecting metal below was in very large mass, there were no depositions of calcareous or magnesian earths upon the metal; it was clean and bright, but never coated. The copper in these experiments was nailed sometimes upon paper, sometimes upon the mere wood, and sometimes upon linen; and the communication was partially interrupted between the external surface and the internal surface by cement; but even one side or junction of a sheet seemed to allow sufficient communication between the moisture on the under surface and the sea-water without, to produce the electrical effect of preservation.

These results upon perfect and imperfect conductors led to another inquiry, important as it relates to the practical application of the principle; namely, as to the extent and nature of the contact or relation between the copper and the preserving metal. I could not produce any protecting action of zinc or iron upon copper through the thinnest stratum of air, or the finest leaf of mica, or of dry paper; but the action of the metals did not seem to be much impaired by the ordinary coating of oxide or rust; nor was it destroyed when the finest bibulous or silver paper, as it is commonly called, was between them, being moistened with sea-water. I made an experiment with different folds of this paper. Pieces of copper were covered with one, two, three, four, five and six folds; and over them were placed pieces of zinc, which were fastened closely to them by thread; each piece of copper so protected was exposed in a vessel of sea-water, so that the folds of paper were all moist.

It was found in the case in which a single leaf of paper was between the zinc and the copper, there was no corrosion of the copper; in the case in which there were two leaves, there was a very slight effect; with three, the corrosion was distinct: and it increased, till with the six folds the protecting power appeared to be lost: and in the case of the single leaf, there was this difference from the result of immediate contact, that there was no deposition of earthy matter. Showing that there was no absolute minute contact of the metals through the moist paper; which was likewise proved by other experiments: for a thin plate of mica, as I have just mentioned, entirely destroyed the protecting effect of zinc: and yet when a hole was made in it, so as to admit a very thin layer of moisture between the zinc and copper, the corrosion of the copper, though not destroyed, was considerably diminished.

The rapid corrosion of iron and zinc, particularly when used to protect metals, only in very small quantities, induced me to try some experiments as to their electro-chemical powers in menstrua out of the contact, or to a certain extent removed from the contact of air, such as might be used for moistening paper under the copper sheathing of ships: the results of these experiments I shall now detail. A small piece of iron was placed in one glass filled with a saturated solution of brine, which contains little or no air; copper, attached by a wire to the iron, was placed in a vessel containing sea-water, which was connected with the brine by moistened tow. The copper did not corrode, and yet the iron was scarcely sensibly acted upon, and that only at the surface of the brine; and a much less effect was produced upon it in many weeks than would have been occasioned by sea-water in as many days.

With zinc and brine in the same kind of connection there was a similar result: but the solution of the zinc was comparatively more rapid than that of the iron, and the copper was rendered more highly negative, as was shown by a slight deposition of earthy matter upon it.

A solution of potassa, or of alkaline substances possessing the electro-positive energy, has nearly the same effect on saline solutions as if they were deprived of air; and when mixed with sea-water impedes the action of metals upon them; but if used in quantity in combinations such as these I have just described, in which iron is the protecting metal, it destroys the result, and renders the iron negative. Thus, if iron and copper in contact, or fastened to each other by wires, be in two vessels of sea-water connected by moist cotton or asbestos, all the various circumstances of protection of the two metals by each other may be exhibited by means of solution of potassa. By adding a few drops of solution of potassa to the water in the glass containing the iron, the negative powers of the copper in the other glass are diminished; so that the deposition of the calcareous and magnesian earths upon it is considerably lessened; by a little more solution of potassa the deposition is destroyed, but still the copper remains clean. The corrosion of the iron, which before was rapid, is now almost at an end; and a few drops more of the solution of potassa produces a perfect equilibrium: so that neither of the metals undergoes any change, and the whole system is in a state of perfect repose. By making the fluid in the glass containing the iron still more alkaline, it no longer corrodes; and the green tint of the sea-water shows that the copper is now the positively electrified metal; and when the solution in the glass containing the iron is strongly alkaline, the copper in the other glass corrodes with great rapidity, and the iron remains in the electro-negative and indestructible state.

I began this paper by some observations upon the nature of the processes by which copper sheathing is destroyed by sea-water, and on the causes by which it is preserved clean, or rendered foul by adhesions of marine vegetables, or animals; I shall conclude it by some further remarks on the same subject, and with some practical inferences and some theoretical elucidations, which naturally arise from the results detailed in the foregoing pages.

The very first experiment that I made on harbour-boats at Portsmouth, proved that a single mass of iron protected fully and entirely many sheets of copper, whether in waves, tides, or currents, so as to make them negatively electrical, and in such a degree as to occasion the deposition of earthy matter upon them; but observations on the effects of the single contact of iron upon a number of sheets of copper, where the junctions and nails were covered with rust, and that had been in a ship for some years, showed that the action was weakened in the case of imperfect connexions by distance, and that the sheets near the protector were more defended than those remote from it. Upon this idea I proposed, that when ships, of which the copper sheathing was old and worn, were to be protected, a greater proportion of iron should be used, and that if possible it should be more distributed. The first experiment of this kind was tried on the Samarang, of 28 guns, in March, 1824, and which had been coppered three years before in India. Cast iron, equal in surface to about $\frac{1}{80}$ of that of the copper was applied in four masses, two near the stern, two on the bows. She made a voyage to Nova Scotia, and returned in January 1825. A false and entirely unfounded statement respecting this vessel was published in most of the newspapers, that the bottom was covered with weeds and barnacles. I was present at Portsmouth soon after she was brought into dock: there was not the smallest weed or shell-fish upon the whole of the bottom from a few feet round the stern protectors to the lead on her bow. Round the stern protectors there was a slight adhesion of rust of iron, and upon this there were some zoophytes of the capillary kind, of an inch and a half or two inches in length, and a number of minute barnacles, both Lepas anatifera and Balanus tintinnabulum. For a considerable space round the protectors, both on the stern and bow, the copper was bright; but the colour became green towards the central parts of the ship; yet even here the rust of verdigrease was a light powder, and only small in quantity, and did not adhere, or come off in scales, and there had been evidently little copper lost in the voyage. That the protectors had not been the cause of the trifling and perfectly insignificant adhesions by any electrical effect, or by occasioning any deposition of earthy matter upon the copper, was evident from this—that the lead on the bow, the part of the ship most exposed to the friction of the water, contained these adhesions in a much more accumulated state than that in which they existed near the stern; and there were none at all on the clean copper round the protectors in the bow; and the slight coating of oxide of iron seems to have been the cause of their appearance.

I had seen this ship come into dock in the spring of 1824, before she was protected, covered with thick green carbonate and submuriate of copper, and with a number of long weeds, principally fuci, and a quantity of zoophytes, adhering to different parts of the bottom; so that this first experiment was highly satisfactory, though made under very unfavourable circumstances.

The only two instances of vessels which have been recently coppered, and which have made voyages furnished with protectors, that I have had an opportunity of examining, are the Elizabeth yacht, belonging to the Earl of Darnley, and the Carnebrea Castle, an Indiaman, belonging to Messrs. Wigram. The yacht was protected by about $\frac{1}{125}$ part of malleable iron placed in two masses in the stern. She had been occasionally employed in sailing, and had been sometimes in harbour, during six months. When I saw her in November she was perfectly clean, and the copper apparently untouched. Lord Darnley informed me that there never had been the slightest adhesion of either weed or shell-fish to her copper, but that a few small barnacles had once appeared on the loose oxide of iron in the neighbourhood of the protectors, which however were immediately and easily washed off. The Carnebrea Castle, a large vessel of upwards of 650 tons, was furnished with four protectors, two on the stern, and two on the bow, equal together to about $\frac{1}{104}$ of the surface of the copper. She had been protected more than twelve months, and had made the voyage to Calcutta and back. She came into the river perfectly bright; and when examined in the dry dock was found entirely free from any adhesion, and offered a beautiful and almost polished surface; and there seemed to be no greater wear of copper than could be accounted for from mechanical causes.

Had these vessels been at rest, I have no doubt there would have been adhesions, at least in Portsmouth or Sheerness harbours, where the water is constantly muddy, and where the smallest irregularity or roughness of surface, from either wear, or the deposition of calcareous matter, or the formation of oxides or carbonates, enable the solid matter floating in the water to rest. There is a ship, the Howe, one of the largest in the Navy, now lying at Sheerness, which was protected by a quantity of cast-iron judged sufficient to save all her copper, nearly fifteen months ago. She has not been examined; but I expect and hope that the bottom will be covered with adhesions, which must be the case if her copper is not corroded; but notwithstanding this, whenever she is wanted for sea, it will only be necessary to put her into dock for a day or two, scrape her copper, and wash it with a small quantity of acidulous water, and she will be in the same state as if newly coppered.

At Liverpool, as I am informed, several ships have been protected, and have returned after voyages to the West Indies, and even to the East Indies. The proportion of protecting metal in all of them has been beyond what I have recommended, $\frac{1}{90}$ to $\frac{1}{70}$; yet two of them have been found perfectly clean, and with the copper untouched after voyages to Demerara; and another nearly in the same state, after two voyages to the same place. Two others have had their bottoms more or less covered with barnacles; but the preservation of the copper has been in all cases judged complete. The iron has been placed along the keel on both sides; and the barnacles, in cases where they have existed, have been generally upon the flat of the bottom; from which it may be concluded, that they adhered either to the oxide of iron, or the calcareous deposits occasioned by the excess of negative electricity.

In the navy the proportion adopted has been only $\frac{1}{250}$ of cast iron, at least for vessels in actual service, and when the object is more cleanness than the preservation of the copper.

It is very difficult to point out the circumstances which have rendered results, such as these mentioned with respect to Liverpool traders, so different under apparently the same circumstances, *i. e.* why ships should exhibit no adhesions or barnacles after two voyages, whilst on another ship, with the same quantity of protection, they should be found after a single voyage.* This may probably depend upon one ship having remained at rest in harbour longer than another, or having been becalmed for a short time in shallow seas, where ova of shell-fish, or young shell-fish existed; or upon oxide of iron being formed, and not washed off, in consequence of calm weather, and which consolidating, was not afterwards separated in the voyage. From what I can learn, however, the chance of a certain degree of foulness, in consequence of the application of the full proportion of protecting metal, will not prevent ship-owners from employing this proportion, as the saving of copper is a very great object; and as long as the copper is sound, no danger is to be apprehended from worms.

It ought to be kept in mind that the larger a ship, the more the experiment is influenced by the imperfect conducting power of the sea-water, and consequently the proportion of protecting metal may be larger without being in excess.

I have mentioned these circumstances because they apply to ships already coppered, and because I have heard that a Liverpool ship, of which it was doubtful whether the copper was in a state such as would enable her to make another voyage to India with security, has, by the application of protectors of $\frac{1}{70}$, made this voyage,† without apparently any wear of her sheathing; and that she is now preparing with the same protectors to make another voyage.

In cases when ships are to be newly sheathed the experiments which have been detailed in the preceding

* The quality of the copper may be another cause.

† The Dorothy.

pages render it likely, that the most advantageous way of applying protection will be under, and not over the copper: the electrical circuit being made in the sea-water passing through the places of junction in the sheets; and in this way every sheet of copper may be provided with nails of iron or zinc, for protecting them to any extent required. By driving the nail into the wood through paper wetted with brine *above* the tarred paper, or felt, or any other substance that may be employed, the incipient action will be diminished; and there is this great advantage, that a considerable part of the metal will, if the protectors are placed in the centre of the sheet, be deposited and re-dissolved: so there is reason to believe that small masses of metal will act for a great length of time. Zinc, in consequence of its forming little or no insoluble compound in brine or sea water, will be preferable to iron for this purpose; and whether this metal or iron be used, the waste will be much less than if the metal was exposed on the outside: and all difficulties with respect to a proper situation in this last case are avoided.

The copper used for sheathing should be the purest that can be obtained; and in being applied to the ship, its surface should be preserved as smooth and equable as possible: and the nails used for fastening should likewise be of pure copper; and a little difference in their thickness and shape will easily compensate for their want of hardness.

In vessels employed for steam navigation the protecting metal can scarcely be in excess;* as the rapid motion of these ships prevents the chance of any ad-

* I have mentioned in the two last communications on this subject some application of the principle; many others will occur. In submarine constructions—to protect wood, as in piles, from the action of

hesions; and the wear of the copper by proper protection is diminished more than two-thirds.

worms, sheathing of copper defended by iron in excess may be used; when the calcareous matter deposited will gradually form a coating of the character and firmness of hard stone.

XIX.

ON THE RELATIONS OF ELECTRICAL AND CHEMICAL CHANGES.*

I. *Introduction.*

A LONG time has elapsed since I read before this Society the Bakerian Lecture on the Chemical Agencies of Electricity. The general laws of decomposition developed in that Paper were immediately illustrated by some practical results, which the Society did me the honour to receive in a very favourable manner; and which, by offering a class of new and powerful agents, led me away for many years into a field of pure chemical inquiry: and it is only lately, and on an occasion which is well known, that I have again taken up the subject of the general principles of electro-chemical action. After a number of new experiments, which I shall have the pleasure of laying before the Society, and notwithstanding the various novel views which have been brought forward in this and in other countries, and the great activity and extension of science, it is peculiarly satisfactory to me to find that I have nothing to alter in the fundamental theory laid down in my original communication; and which, after a lapse of twenty years, has continued, as it was in the beginning, the guide and foundation of all my researches.

I am the more inclined to bring forward these new

* [From Phil. Trans. for 1826. Read before the Royal Society, June 8, 1826.]

labours at the present moment, though they are far from being in a finished state, because the discovery of Œrsted and that of Morichini, illustrated by some late ingenious inquiries, connect the electro-chemical changes with entirely new classes of facts, and induce a hope that many of the complicated phenonena of corpuscular changes, now obscure, will ultimately be found to depend upon the same causes, and to be governed by the same laws; and that the simplicity of our scientific arrangements will increase with every advance in the true knowledge of nature.

II. *Some Historical Details.*

As I am not acquainted with any work in which full and accurate statements on the origin and progress of electro-chemical science are to be found, and as some very erroneous statements have been published abroad and repeated in this country, I shall take the liberty of laying before the Society a short historical sketch on this subject; which is the more wanted, as the journal in which the early discoveries were registered has long been discontinued, and is now little known or referred to.

As there are historians of chemistry and astronomy who date the origin of these sciences from antediluvian times, so there are not wanting persons who imagine the origin of electro-chemical science before the discovery of the pile of Volta; and Ritter and Winterl have been quoted* amongst other persons as having imagined, or anticipated the relation between electrical powers and chemical affinities, before the period of this great invention. But whoever will read with attention Ritter's "Evidence that the galvanic action exists in organized Nature,"† and Winterl's "Prolusiones ad Chemiam

* Œrsted, translated by Marcel, 1813.

† Jena, 1800.

Sæculi decimi noni," will find nothing to justify this opinion. Ritter's work contains some very ingenious and original experiments on the formation and powers of single galvanic circles; and Winterl's some bold, though loose speculative views* upon primary causes of chemical phenomena: and in the obscurity of the language and metaphysics of both these gentlemen, it is difficult to say what may not be found. In the ingenious, though wild views, and often inexact experiments of Ritter, there are more hints which may be considered as applying to electro-magnetism than to electro-chemistry; and Winterl's miraculous "*Andronia*" might, with as much propriety, be considered as a type of all the chemical substances that have been since discovered, as his view of the antagonist powers, the acid and basic, can be regarded as an anticipation of the electro-chemical theory. The queries of Newton at the end of his "Optics" contain more grand and speculative views that might be brought to bear upon this question than any found in the works of modern electricians;† but it is very unjust to the experimentalists who, by the laborious application of new instruments, have discovered novel facts and analogies, to refer them to any such sup-

* As a specimen of the Prolusiones, I shall give a few articles from the Index, which will show the character of the work. Prolusiones, p. 256, et seq.

256. "Adamas est Andronia.

260. "Andronia cum Plumbo creat Barytam cum Ferro Chalybem.

262. "Carbo est acidus cum Atmosphæra basica.

263. "Chromium non est nisi Calx Magnesii acida.

—. "Cuprum cum Andronià coalescit in Molybdænum.

268. "Scintilla electrica formatur à Principiis Conductorem primum et secundum animantibus, ac inter se concurrentibus; est gravis, habet effectum electricitati contrarium."

† See the eloquent observations of Mr. Chevenix on the subject of Winterl's Theory, Annales de Chim. vol. 50, 2 Cap. 175.

positions as, "that all attractions, chemical,* electrical, magnetic, and gravitative, may depend upon the same cause;" or to still looser expressions, in which different words are used and applied to the same ideas, and in which all the phenomena of nature are supposed to depend on the dynamic system, or the equilibrium and opposition of antagonist powers.

The true origin of all that has been done in electro-chemical science was the accidental discovery of MM. Nicholson and Carlisle, of the decomposition of water by the pile of Volta, April 30, 1800.† These gentlemen immediately added to this capital fact, the knowledge of the decomposition of certain metallic solutions, and the circumstance of the separation of alkali on the negative plates of the apparatus. Mr. Cruickshank, in pursuing their experiments, added to them many important new results, such as the decomposition of muriates of magnesia, soda and ammonia, by the pile; and that alkaline matter always appeared at the negative,

* In the Systême Universelle of M. Azais, not only are all the phenomena of nature referred to the same cause, but specific reasonings upon the mode of its operation given. In this work, published in 1810, not only is the identity of magnetism and electricity insisted on, but an attempt is made to explain the manner in which the two electrical fluids produce the magnetic phenomena, p. 239, vol. i. "Ainsi ces deux ordres de phénomènes sont tres resemblans. Repetons que toutes leurs différences résultent uniquement de ce que les deux fluides sont moins intenses lorsq'ils produisent les phénomènes du Magnétisme que lorsq'ils produisent les phénomènes du Galvanisme, &c." It requires only the same principle as that censured in the text to refer to this author the discovery of Œrsted and the speculations of Ampère. M. Azais, in his "fluides mineure et majeure," finds all the causes of the acid and alkaline properties of bodies:—slow combinations, the heat produced, and all the phenomena of chemical change; and his reasonings are often very ingenious.

† Nicholson's Journal, vol. iv. p. 183.

and acid at the positive pole:* and Dr. Henry about the same time made some unsuccessful attempts to decompose potassa in solution by the pile, and confirmed the general conclusions of MM. Nicholson, Carlisle and Cruickshank. In the month of September in this year, I published my first paper on the subject of Galvanic Electricity, in Nicholson's Journal, which was followed by six others:† the last of which appeared in January, 1801. In these papers I showed that oxygen and hydrogen were evolved from separate portions of water, though vegetable and even animal substances intervened; and conceiving that all decompositions might be polar,‡ I electrised different compounds at the different extremities, and found that sulphur and metallic substances appeared at the negative pole, and oxygen and azote at the positive pole, though the bodies furnishing them were separate from each other. In the same series of papers I established the intimate connexion between the electrical effects and the chemical changes going on in the pile, and drew the conclusion of the dependence of one upon the other. In 1802 I proved that galvanic combinations might be formed from single metals, or charcoal and different fluids chiefly acid and alkaline, and that the side or pole of the conducting substance in contact with the alkali was positive, and that in contact with the acid, negative; and in the same year I published, that when two separate portions of water, connected by moist bladder or muscular fibre, were electrised, nitro-muriatic acid appeared at the positive, and fixed alkali at the negative pole.‡ In the same year Dr. Wollaston placed the iden-

* Nicholson's Journal, vol. iv. p. 190.

† Ibid. pp. 275, 326, 337, 394, 380. [Vide Vol. II. p. 139 et seq.]

‡ Journal of the Royal Institution, 1802, First Series. [Vol. II. p. 202.]

tity of the cause of galvanism and electricity, which had been always maintained by Volta, out of all doubt, by some very decisive experiments.

In 1804, MM. Hisinger and Berzelius stated that neutro-saline solutions were decomposed by electricity, and the acid matter separated at the positive, and the alkaline matter at the negative poles; and they asserted that in this way muriate of lime might be decomposed; and drew the conclusion that nascent hydrogen was not, as had been generally believed, the cause of the appearance of metals from metallic solutions. These valuable observations ought to have explained distinctly the source of the appearance of acid and alkaline substances at the two extremities of the pile, yet the paper was never translated into English, nor at all attended to; and one of their facts was contradicted by the assertion of, generally, a very accurate observer, Mr. Cruickshank, who in his early experiments mentioned that he had not been able to decompose muriate of lime in the circuit.

In 1805 various statements were made, both in Italy and England, respecting the generation of muriatic acid and fixed alkali from pure water. The fact was asserted by MM. Pacchioni and Peele, and denied by Dr. Wollaston, M. Biot, and the Galvanic Society at Paris.* Mr. Sylvester, who conducted his experiments

* Some writers have very incorrectly referred the origin of these researches to the observations of Hisinger and Berzelius; Annales de Chim. Vol. li. 1 Cap. page 167; but these observations were never quoted by any writer of the day on the pretended production of muriatic acid and alkali; and I was not acquainted with them till after my fundamental experiments were finished; and when, in drawing up an account of them, I looked back through the whole series of periodical publications to find accounts of experiments bearing upon the same question, and I believe I first directed the public attention to the value

with some care, stated that if two separate portions of water were electrised out of the contact of substances containing alkaline or acid matter, acid and alkali were generated; so that up to this time the question, whether these substances were liberated from their combinations, or formed from their elements by electricity, could not be considered as decided: a circumstance not so much to be wondered at, when the novel and extraordinary nature of the whole class of galvanic phenomena is considered.

It was in the beginning of 1806* that I attempted the solution of this question; and after some months' labour I presented to the Society the dissertation, to which I have referred in the beginning of this lecture. Finding that acid and alkaline substances, even when existing in the most solid combinations, or in the smallest proportion in the hardest bodies, were elicited by voltaic electricity, I established that they were the results of decomposition, and not of composition or generation; and referring to my experiments of 1800, and 1801 and 1802, and to a number of new facts, which showed that inflammable substances and oxygen, alkalies and acids, and oxidable and noble metals, were

of those researches. Whoever will take the trouble to read the Bakerian Lecture for 1806, will be convinced of the gradual development of the whole subject from the investigation respecting the pretended formation of muriatic acid and fixed alkali. [M. Berzelius in a controversial paper, a translation of which was published in the 8th vol. of Ann. of Philosophy, 1816, entitled "Comparison of the old and new theories respecting the nature of oxy-muriatic acid," remarks "In the year 1806, we were not aware of the reducing power of Davy's electrical piles and troughs." This observation incidentally made may be considered an acknowledgment of M. Berzelius' ignorance of the power in question at that time, and that he had not viewed in their consequences the results of his earlier experiments.]

* Phil. Trans. 1807.

in electrical relations of positive and negative, I drew the conclusion, *that the combinations and decompositions by electricity were referable to the law of electrical attractions and repulsions*, and advanced the hypothesis, "*that chemical and electrical attraction were produced by the same cause, acting in one case on particles, in the other on masses;" and that the same property, under different modifications, was the cause of all the phenomena exhibited by different voltaic combinations.*

Believing that our philosophical systems are exceedingly imperfect, I never attached much importance to this hypothesis; but having formed it after a copious induction of facts, and having gained immediately by the application of it a number of practical results, and considering myself as much the author of it as I was of the decomposition of the alkalies, and having developed it in an elementary work, as far as the present state of chemistry seemed to allow, I have never criticised or examined the manner in which different authors have adopted or explained it,—contented, if in the hands of others it assisted the arrangements of chemistry or mineralogy, or became an instrument of discovery. And having now given what I believe to be a faithful sketch of its origin, I shall not enter into an examination of those works which have induced me to make this sketch, and which contain partial or loose statements on the subject, and which refer the origin of electro-chemistry to Germany, Sweden and France, rather than to Italy and England, and which attribute some of the views of the science, which I first developed, to philosophers who have never made any claim of the kind, and who never could have made any, as their works on the subject were published many years after 1806.

III. *On the Modes adopted for detecting the Electrical States of Bodies, and Definitions of Terms.*

That the statements made in the following sections may be more distinct, I shall say a few words of the mode in which the different conditions of electrical action were ascertained, and describe the manner in which I have used the terms which have been adopted in electro-chemical science.

In determining the nature of the electrical action in what may be called the closed circle, or the combinations in which, according to the language used on the Continent, electrical currents exist, I have employed instruments constructed upon the same principles as the galvanometer of Professor Cumming, or the multiplier of Professor Schweigger. Silver wire, covered with silk, about $\frac{1}{70}$ of an inch in diameter, was folded round a small wooden frame, so as to fill a narrow deep groove: the two extreme wires were parallel, and the convolutions as nearly as possible in the same perpendicular: a small tube containing a filament of silk was passed through the centre of the convolutions of wires, to which a delicate magnetised needle was suspended; which, when the apparatus was properly disposed, rested with its north pole between the two extremities of the wires. This instrument, which contained sixty circumvolutions of wire, was found sufficiently delicate for most purposes of experiment; but in a few instances, in which very weak electricities were to be determined, I used another apparatus, in which the same kind of wire was fastened, in concentric circles, round two portions of glass tube, in such a manner that radii from the inner circle would have passed through all the wires, and in which increased mobility was given to the system by two needles exterior to it and connected with it, placed one above, the other below the central needle, with their poles in the same directions, but opposite to those of the central needle, which was so magnetised that its directive power was neutralised by the power of the other two needles.*

To illustrate the operation of these apparatus, I shall state, that when the lower terminating wire, which was to the left, or east of the north pole, was connected with a piece of zinc, and the upper one with a piece of platinum, both being in common water, the deviation of the central needle was eight or ten degrees, the south pole turning to the east or left hand; which may be considered as indicating that the current of electricity was from the platinum to the zinc through the wire, and that the surface of the zinc in the fluid was positive with respect to the opposite surface of platinum; and in using the terms positive and negative, I beg to be understood as applying them to the metallic surfaces in contact with the fluid.

For determining weak electricities of charge, or as it is sometimes called, of tension, I used Volta's condenser connected with Bennet's electrometer, and sometimes with one constructed on the principle of Behrens, consisting of an insulated gold leaf, or what I found better, a silk filament, made conducting by impalpable charcoal powder, to receive the charge, placed between the poles of a dry pile consisting of 400 circles of silver and gold foil, of the third of an inch in diameter, or 50 of zinc and silver of the same size, with paper intervening; the attraction of the gold leaf or the filament, either to the positive or negative pole, indicates the nature of the charge: and, as in cases of electro-chemical action

* This arrangement differs from that of M. Nobile only by a duplication of effect.

there are always two corresponding opposite states, I considered the part of the system which touched the conductor as possessing the same electrical state with that exhibited by the leaf. I have never however put much dependence upon indications given by this instrument, unless they were confirmed by other results; having found them very uncertain and influenced by the state of the condenser and the atmosphere.

IV. *On the Electrical and Chemical Effects exhibited by Combinations containing single Metals and one Fluid.*

I know of no class of phenomena more calculated to give just views of the nature of electro-chemical action than those presented by single metals and fluids; and as their results are, with one or two exceptions, entirely new, I shall describe them with some degree of minuteness. When two pieces of the same polished copper, connected with the platinum wires of the multiplier, were introduced at the same time into the same solution of hydro-sulphuret of potassa, there was no action; but if they were introduced in succession, there was a distinct and often, if the interval of time was considerable, a violent electrical effect — the piece of metal first plunged in being negative, and the other positive.

This result depends upon the circumstance of the production of a new combination, which is negative with respect to the metal; for after the formation of the sulphuret of copper, the plate of copper that has been first plunged into the solution exhibits the same negative state with respect to polished copper, whether introduced into saline solutions, or alkaline or acid menstrua. The electrical effect therefore does not depend on so simple a condition as would at first appear, and it may be in fact referred to the combination containing two metallic substances and one fluid.

The grey sulphuret of copper is negative, in solutions of hydro-sulphuret, to clean copper, and the superficial coating has apparently similar electrical powers to this substance.

Copper, in the state of protoxide, is negative, not only with respect to metallic copper, but likewise with respect to the sulphuret; a circumstance which explains many singular and apparently anomalous circumstances with respect to the action of hydro-sulphuret on copper. I have often found the order which I have mentioned, of metallic copper being positive with respect to copper that had been a few seconds in solution of hydro-sulphuret, reversed in a singular and capricious way; but on investigating the cause, I found that the copper was tarnished; and on heating any kind of polished copper strongly, so as to produce a thin coating of oxide any where on its surface, it became strongly negative to copper plunged in solution of hydro-sulphuret: the same effect was produced by the action of acids.

There are some singular circumstances connected with the violent and intense chemical action of copper on solutions of hydro-sulphurets, which are worthy of being described. When a piece of copper connected with the multiplier has been for a minute in strong solution of hydro-sulphuret of potassa, on introducing a piece of polished copper connected with the other wire, there is often a violent and momentary negative charge communicated to it, which sends the needle through a whole revolution: it then oscillates, and almost immediately returns, and takes the direction which indicates that the piece first plunged in is negative. This effect continues for some minutes, then becomes weaker; at last the two sides are in equilibrium, and the piece which was first plunged in now becomes positive with respect to the other. The first described of these effects seem to depend upon the discharge, by the clean copper, of the negative electricity accumulated by the contact of the plate first plunged in, before the relative states produced by the metallic contact and the regular currents occur; and the second, to the detaching or peeling off of the coat of sulphuret, which has the effect of exposing a clean surface, and which effect is probably occasioned by the oxidation of the positive side of the plate.

There are few electrical actions more intense than those produced by the operation of hydro-sulphurets on copper in these different circumstances; so much so, that I have constructed a voltaic battery which decomposed water, by six combinations, consisting merely of thin slips of copper, of which one-half had been exposed to the solution about a minute before the other half: of course, the oxidating surface was on the side of the clean or latest exposed metal.

With lead, and alloys of tin and lead and iron, there are the same phenomena, but much feebler electrical action, the metallic surface, which is first introduced, being the negative surface; and the principles of this kind of action are precisely the same as those of copper and hydro-sulphurets.

Zinc, platinum, and metals which have no chemical action on solutions of hydro-sulphurets, produce no phenomena of this kind; silver and palladium, which act powerfully with these menstrua, produce very decided effects; but the compounds they form in them being positive with respect to the pure metals, the phenomena are the reverse of those offered by the more oxidable metals: the surface plunged first into the solution is the positive surface, and it retains this relation in alkaline, acid, and saline solutions, presenting peculiarities dependent upon the change of surface, which I shall refer to again hereafter.

The production of electrical currents by single metals and single fluids, though most distinct in the cases I have just named, yet occurs generally whenever new substances, which can adhere to the metals, are produced in chemical action. Thus, in acid solutions of a certain strength, pieces of the same zinc, tin, iron, and copper, exhibit similar phenomena; the surface first plunged into the acid being tarnished, or retaining a slight coat of oxide, is negative to the metal plunged in afterwards, and the relation is sustained in saline or alkaline solutions. The same effect is caused by producing a coat of oxide by heat on the surface, or even by applying it artificially. The oxidated surface is negative with respect to the other.

Zinc, which dissolves in a strong solution of potassa, giving off hydrogen copiously, exhibits exactly the same phenomena in this solution; the tarnished metal, or that first introduced, being negative with respect to the other. Tin, likewise, in solution of potassa, having been introduced long enough to have tarnished, is strongly negative with respect to polished tin.

Even the noble metals obey the same law. Silver, that has been tarnished by the action of nitric acid, is negative to polished silver in diluted acid; and gold and platinum, that have been acted on by aqua regia, are negative in that acid to the clean metals.

The intimate connexion displayed in all these cases between the chemical and electrical phenomena, becomes still more remarkable when the nature of the changes taking place in circles of this kind is considered.

Oxygen, which may be considered as negative with respect to all the metals, and sulphur, which is negative with respect to the oxidable metals, by their combinations with metals respectively positive to them, produce compounds negative with regard to those metals. And in the chemical changes, the results are such as must ultimately restore the equilibrium, hydrogen or sulphuretted hydrogen passing to the negative side, and oxygen to the positive side; so that the oxides are revived; and not only is the equilibrium restored, but the poles sometimes changed. Thus tin that has tarnished in acid, remains for some time negative in solution of alkali, but gradually as the oxide upon it is revived by the hydrogen determined to this surface, it loses its negative power; and the other surface, now tarnished by the action of the alkali, gains this power, whilst the opposite surface becomes positive.

V. *Of electrical Combinations, consisting of two imperfect, and one perfect Conductor; or two Fluids and a Metal, or Charcoal.*

To understand clearly the nature of the action in this kind of electrical combination, it is necessary to consider the nature of imperfect conducting bodies, water, or saline solutions. These bodies may be regarded as having the same relations to electricities of very low intensity, that elastic fluids have to the electricities of glass, sealing wax, or the common machine. They communicate the electrical polarities of the metals, but do not appear capable of receiving such polarities, or at least of retaining them; and the electrical equilibrium, when broken in them, seems to be rapidly restored by a new arrangement or attraction of certain of their elements. For instance, if we introduce the positive and negative poles from a very powerful Voltaic battery into the extremities of a basin filled with solution of muriate of lime, and place in the circuit different wires of platinum, every wire will possess a positive and negative pole, and there will be no division of the fluid into two parts—one positive, the other negative; and when the two wires are withdrawn, they alone having been used, the electrical appearances immediately cease; and metallic wires, unconnected with the battery, made to occupy their places, exhibit no electrical phenomena: and in all experiments of this kind, the well known phenomena of the development of chlorine and oxygen and acid matter at the positive, and hydrogen, alkaline matter, &c. at the negative pole, take place.

Acid and alkaline matters, when perfectly dry and non-conducting, become, on contact, negative and positive; as I have shown is the case with oxalic acid and lime; but this effect is similar to that of glass and silk, and the result is a common electricity of tension. And when acids and alkalies combine, their union being apparently the result of the same attractive powers acting on the particles, which would produce their electrical relations as masses, they exhibit no phenomena of electro-motion; and such phenomena, when they occur in combinations in which acids and alkalies unite, always depend upon the contact of the metal with the acid and alkaline matter, change of temperature, evaporation, &c. and never on the combination of the acid and alkali.*

* [The cause of voltaic electricity, as well as of every other kind of electricity, is still a quæstio vexata; and is long likely to remain such.

As a different opinion has been lately started, on high authority,* I shall give the proofs of the truth of this, my early view, which appear to me of the strictest demonstrative nature.

A solution of nitre, which is a substance neutral to the contact of noble metals, was introduced into a glass cup containing a plate of platinum, connected with the multiplier; pure concentrated nitric acid was placed in another cup, in which there was another plate of platinum joined to the other wire of the multiplier, and the connection was made by a piece of asbestus wetted in a solution of nitre. At the moment of contact, the needle indicated a strong electrical action, negative on the plate plunged in the acid, and which occasioned a permanent deviation of about 60°.

This arrangement was removed from the multiplier, and another substituted for it, in which strong solution of potassa occupied the place of the nitric acid, being in contact with platinum in one cup, and solution of nitre in the other, with the same communications. The deviation was now much weaker, about 10 degrees,—and the platinum in the solution of potassa, was positive.

The nitric acid and the solution of potassa were now connected in the arrangement by a piece of clean asbestus, moistened in a concentrated solution of nitre; the deviation of the needle was to about 65°. In this instance, there was no chemical action of the fluids on

The further inquiry is extended, the more mysterious the subject appears, not unanalogous, in this respect, to the principle or cause of life. The view which the author took of electrical excitation, he considered merely as most probable—one of many hypotheses, which might be formed respecting it; this should be kept in mind, and further that he attached no importance to his hypothetical view, excepting in connexion with facts, and as a guide to further experiments.]

* That of M. Becquerel.

each other; for they had no tendency to mix rapidly with the solution of nitre, which being of less specific gravity than either of the other solutions, remained in the asbestus; and there was no effect, beyond that of the metallic contact of the platinum with acid and alkali.

A piece of asbestus, of nearly the same size with the other, but dry, was now substituted for the moist asbestus, so that the acid and alkali combined by capillary attraction producing heat;—at first, the deviation was rather less, than in the former instance; but as soon as the combination was complete, the needle stood exactly at the same point,—proving that no electricity was developed by the combination, any more than by the indirect communication of the acid and the alkali.

After trying the effects of the contact of fluid acid upon platinum, by the arrangement with solution of nitre, and finding that oxalic acid was the acid among the powerful ones which produced the slightest deviation of the needle, or the smallest negative effect, I employed this acid and solution of potassa, exactly in the same manner as the nitric acid in the experiment just detailed; as the joint action of the acid and alkali on the platinum, was only to produce a deviation of 7 or 8 degrees, it might be suspected that any electrical action produced by combination, might be more easily manifested; but no such effect occurred: and whether the communication was made by combination, through dry asbestus, or through asbestus, wetted in a saline solution, the effect was precisely the same.

Again,—the two surfaces of platinum were placed in contact with strong solutions of nitre, and the communication made between them by solution of potassa and nitric acid; there was no electrical action, though the chemical combination was intense. But when the fluids were mixed, so that a little acid touched one plate of platinum, and a little alkali the other, electro-motion immediately began; and in using muriatic acid and solution of ammonia, which, being lighter than the saline solutions, very soon came in contact with the platinum, the effect commenced almost immediately, and continued for some time to increase.

Again, — I placed pieces of paper coloured with litmus and turmeric, and moistened in solutions of nitre, upon two surfaces of platinum connected with the multiplier; they were covered with a stratum of porcelain clay wetted with the same solution, a stratum of clay moistened with muriatic acid was placed on one plate, and a stratum moistened with solution of ammonia above on the other, so as to make a contact in which there should be action upon a large surface without direct communication with the metals. In several experiments of this kind there was no electro-motion; and whenever it was perceived it was found that either the acid, or the alkali, or both, had penetrated through the clay, and touched the metals so as to change considerably the colour of the papers, which were placed as indications of the correctness of the experiment.

Having brought forward what appear to me decided proofs on this subject, I shall now proceed to investigate the operation of the metals and fluids in combinations containing two of the latter substances. At first I was surprised to find that platinum acted so powerfully with nitric acid, which undergoes no chemical change by contact with it, and suspecting that it might arise from the presence of minute portions of muriatic acid or muriatic salts, I took great pains to exclude these substances by washing the platinum in distilled water, not touching it with the hands, &c., but when the conditions were those of perfectly clean and pure platinum and perfectly pure nitric acid, the phenomena were the same. Similar reasonings may be applied to solutions of potassa, soda, &c., which do not chemically alter platinum by contact, and yet render it positively electrical with respect to platinum in water or saline solutions. It must however be called to mind that the oxygen in nitric acid, and the metals in the alkalies, have attractions of a very decided kind for platinum; and in taking the scale of electro-negative bodies, solutions of chlorine, or nitro-muriatic acid, produce a more powerful electrical effect on platinum than nitric acid, nitric acid than muriatic, and muriatic than sulphuric.

When platinum is brought in contact with an acid, the pole touching the acid is negative, the opposite pole is positive, as I have found by the condensing electrometer; and the reverse is the case when it touches an alkali, so that the circulation of the electricity is from the metal to the alkali, and from the acid to the metal.

Rhodium, iridium, and gold, act in combinations consisting of acid and alkali, on which they have no chemical effect, exactly like platinum; the surface of the metal in the solution of alkali being positive, that in the solution of the acid, negative. With silver and palladium the electricity is greater, particularly if nitric acid is used; and with charcoal and oxidable metals, there is the same general result, the action being in general exalted in proportion as the chemical attractions are stronger, provided there are no interfering circumstances: and in combinations of this kind nitro-muriatic acid is more active than nitric, and the order is after, nitric, nitrous sulphuric, phosphoric, vegetable acids, sulphurous, prussic, sulphuretted hydrogen, and, with the alkalies, potassa, soda, baryta, ammonia, and so on.

It is always to be understood that strong or concentrated solutions of acids and alkalies are employed; for in cases where the quantity of acid or alkaline matter is very small and the chemical action of the metals strong, there is sometimes a different order. Thus zinc and tin tarnished immediately even in a weak solution of potassa, and, so tarnished, they are negative to the same metals in weak solutions of muriatic or sulphuric acid; but in experiments of this kind it is easy to determine the true circumstances by changing the poles; the negative side, when the energies of the alkali and acid are weak, will be determined by the tarnish or coat of oxide formed.

Solutions of sulphurets act in these combinations like alkali, with circumstances depending upon the formation of new compounds according to the law explained in the last section. In combinations, of which the elements are hydro-sulphuret and acid, the metal in the hydro-sulphuretted solution is positive, and that in the acid negative; but with alkalies and hydro-sulphurets, and zinc and tin, the metal in the solution of alkali is positive, and that in the solution of hydro-sulphuret negative: with silver and palladium the opposite order occurs, and with copper there is nearly a balance of powers, or changes of power, dependent upon the circumstances detailed in the last section.

When, in electrical combinations containing one metal, water, or a neutro-saline solution is in one of the cups, and alkali or acid in another, the result is usually such as might be anticipated,—the side of the metal in the alkali is positive, that in the acid negative, and that in the neutro-saline solution in the opposite state. There are however certain neutro-saline solutions, which, when they contain oxygen or common air, act upon oxidable metals, and such have a power or energy of their own; thus zinc, and tin, and copper in solution of common salt, are positive to the same metals in distilled water; and the surfaces of the same metals in weak muriatic acid are positive with respect to the surfaces in water or saline solutions. In combinations, in which weak and strong solutions of acids or of alkalies are the two fluids, both being of the same kind, the electrical action is usually feeble; but the surface in the strongest alkali is most positive, and in the acids the result usually depends upon the nature of the solution; if oxide is formed and deposited, the strongest acid is negative with respect to the diluted one.

The chemical changes produced in combinations of this kind, are best observed in cases where the metals undergo no change; for instance with platinum, diluted sulphuric acid, and solution of potassa. In this combination, hydrogen soon appears on the platinum in the acid, and a very small quantity of gas, which is probably oxygen, on the platinum in contact with the alkali; and that the acid tends to circulate towards the negative surface, and the alkali towards the positive, is shown by the circumstance of the rapid neutralization of the two menstrua, though separated by asbestus moistened in distilled water.

VI. *Of Combinations consisting of two Conductors of the more perfect Class, and one Fluid.*

The order in which metallic bodies exhibit electricities on contact, as is well known, is intimately connected with their relative oxidability, the most oxidable metal being positive with respect to all those below it. This law extends likewise to the newly discovered bases of the alkalies and earths. Potassium and sodium, as I have found by bringing them in contact with zinc in a concentrated solution of the alkali, are apparently as much positive with respect to this body, as zinc is with respect to platinum and gold.

There is not however any inherent and specific property in each metal which gives it the electrical character; it depends upon its peculiar state — on that form of aggregation which fits it for chemical change. Thus, zinc in amalgamation with mercury is positive with respect to pure zinc, and the amalgam of tin is in the same state with regard to tin; and the metals of the fixed alkalies in amalgam give the highest positive energy to a mass of mercury some thousands of times their weight.

In general, the electricities developed by metallic contact are of a stronger kind than those resulting from the contact of metals with fluids, so that they are not capable of being changed by them. For instance: zinc in acid is positive with respect to all other metals below it in degree of oxidability, though they are placed in alkalies or solutions of sulphurets: there are however exceptions; for instance, with regard to tin, which, when in a strong solution of potassa is positive to zinc in an acid solution; and with respect to iron, which, though positive with regard to copper in all acid or neutro-saline fluids, is negative to it in solution of sulphurets or of alkalies. The electro-motion in these instances produced by the contact of the fluids prevailing over that produced by the contact of the metals.

And knowing the energies of the acid and alkaline fluids, it is easy to apply them so as to diminish or enhance the electrical effects developed by metallic contact.

If, for instance, in a combination containing zinc and platinum, we use two fluids, and place the acid in contact with the zinc, and the alkali with the platinum, the effect will be exceedingly feeble compared with that produced if the order be reversed, and the zinc be in contact with the alkali, and the platinum with the acid.

The chemical changes taking place in combinations of this kind are always such as tend to restore the equilibrium, the hydrogen and the alkaline body always passing to the negative, and oxygen and the acid to the positive metal.

There is no instance of continued electro-motion except in cases where chemical changes can take place, for even De Luc's or Zamboni's columns do not act when quite dry, and the silver in combinations of this kind, when the negative metal is gold, is uniformly found tarnished: for the exhibition of electricities of tension, however, a very slight chemical action is sufficient, as the quantity of electricity required to give repulsion to light bodies is exceedingly small; but to form electro-magnetic combinations the chemical agents must be of an energetic kind.

As most of the fluids which act powerfully in voltaic combinations contain water, or oxygen and hydrogen, it has been suspected that these principles were essential to the effect;* this however does not seem to be

* [It has been supposed that the author once entertained this opinion and expressed it, in his Elements of Chemical Philosophy, page 123, where he says "There are no fluids known, except such as contain water, which are capable of being made the medium of connection

the case, for I found zinc and platinum formed powerful electro-motive circles in fused litharge and fused oxy-chlorate of potassa, which are not known to contain water; and I have little doubt that similar effects would be produced by other fused salts containing only acid and alkaline matter.

It may elucidate this part of the subject, which must at best be obscure, to take a view of the changes occurring in one of the simplest voltaic combinations, that consisting of zinc, platinum, and solution of sulphate of soda. It is a fact that zinc and platinum become electrical by contact, the zinc positive, the platinum negative; and the two kinds of electricity are apparently most intense at the surfaces where they are in contact with the fluid, which is too imperfect a conductor to allow them to neutralize or destroy each other: they consequently exert their attractive and repellent powers upon the elements of the menstruum; acid and oxygen circulate to the surface of zinc, which in consequence is dissolved, and alkali and hydrogen to the surface of platinum, of which the hydrogen is disengaged, and the equilibrium broken by the contact of the metals is restored by the chemical changes; so that a constant circulation, or a current of electricity, takes place, the power of the combination becoming feebler in proportion as the solution is decomposed, and acid accumulated round its positive, and alkali round its negative surface.

In cases where acids or acid solutions alone are used, the destruction of one or both surfaces, with the transfer of hydrogen or oxygen, seems to produce the same

between the metals or metal of the voltaic apparatus;" but, as it appears from the context, that he was speaking of the apparatus as it was commonly used, the supposition is not warranted.]

effect; and the inactivity of single circles or voltaic piles, in which pure water is used or saline solutions freed from air, seems to show that the destruction of the surface of the oxidable metal is one of the conditions of continued electrical action ; and the cessation of the power of De Luc's or Zamboni's piles, is always connected with the tarnish of the imperfect metal employed in them.

Having published many years ago tables of the electro-chemical relations of metals, which have been copied into many elementary books, I think it proper to give them here in a corrected form with some ad ditions, and the differences dependent upon the nature of the menstruum. The metal mentioned first is positive to all those below it in the scale.

With common acids.—Potassium and its amalgams; barium and its amalgams; amalgam of zinc; zinc; amalgam of ammonium (?); cadmium, tin, iron, bismuth, antimony (?), lead, copper, silver, palladium, tellurium, gold, charcoal, platinum, iridium, rhodium.

With alkaline solutions.—The alkaline metals and their amalgams: zinc, tin, lead, copper, iron, silver, palladium, gold, platinum, &c.

With solutions of hydro-sulphurets.—Zinc, tin, copper, iron, bismuth, silver, platinum, palladium, gold, charcoal.

VII. *On the Accumulation of Electricity, and the chemical Changes it occasions in Voltaic Arrangements.*

In the view of electro-motion adopted by the illustrious inventor of the pile, the metals were considered as the *only* agents which, in proportion to their surface and their number, occasioned the constant circulation of a certain quantity of electricity through the fluids, or the connecting wires in the pile; and the chemical changes occurring in these fluids were considered as mere results, and not necessarily connected with the circulation. The inactivity of combinations where no chemical changes occur, is sufficiently hostile to this view; but an examination of some of the circumstances of the construction of compound electrical combinations, will bring this hypothesis, and that which I have ventured to adopt, more distinctly into comparison.

Let a piece of zinc and a piece of platinum, both in glasses filled with a solution of nitrate of potassa, be connected through the multiplier, and let the glasses be joined by asbestus moistened with the same fluid; the needle will mark electrical action: let the two glasses now be joined by an arc composed of zinc and platinum, in such a manner that the order is voltaic, *i.e.* that the zinc is opposite to the platinum, in the original combination—the effect will be increased. Now let an arc of pure zinc be introduced; the effect will be less than with the double arc, but superior to that with the asbestus, and the pole of the zinc opposite the platinum will oxidate, and that opposite the zinc will give off hydrogen. Let arcs of other metals be substituted for the zinc; for instance, of tin, of iron, of copper, of silver, of tellurium: the electrical effects will diminish with the oxidability of the metal; and with tellurium, which does not oxidate at the positive pole of a voltaic battery, they will be destroyed; and the case is the same with rhodium, palladium and platinum. That the effect does not depend upon any circumstance connected with conducting power is evident; for charcoal, which is a very imperfect conductor, acts like an oxidable metal; and a very fine wire of platinum, terminated by a small piece of oxidable metal, acts more efficiently when the oxidable metal is opposite the negative pole than if the whole chain had been composed of oxidable metal; but entirely destroys the effect when the oxidable metal is opposite the positive pole.

If the contact of the metals only was necessary for continued electro-motion, these results, in which a simple homogeneous chain is interposed between the fluids, would be impossible; but they are a necessary consequence of the electro-chemical theory, in which the destruction of the positive surface by the chemical negative agent is regarded as a necessary condition; and platinum and tellurium acted like zinc, when their surfaces opposite to the platinum were plunged into diluted nitro-muriatic acid.

If two, three, or four glasses are used, and two, three, or four arcs of platinum and zinc, the extreme metals of which are connected through the multiplier, a piece of platinum used instead of one of the arcs will not now entirely destroy the electro-motive effect; it will be diminished as if one arc had been removed. The two will act as a single combination; the three as two arcs, and the four as three, and of course in a voltaic combination of 100 arcs, a single piece of platinum substituted for any one of the arcs, will diminish the power of the apparatus only $\frac{1}{100}$ part.

In attempting to protect copper by zinc, in a separate vessel, from the action of sea-water, I found that when the two vessels were connected by moist tow or vegetable substances, or by a wire (even though fine) of any oxidable metal, the protection was complete: but when even a thick wire of platinum was employed, the copper, though in immediate contact with the zinc, became corroded. After the experiment had continued several days, the surface of the platinum opposite to the copper was found tarnished, as if it had been slightly acted upon by the chlorine combined in the sea-water; but this effect had been too feeble to be connected with any sensible degree of electrical polarity in the platinum.

This result, with those mentioned in the preceding pages, seems to show that there can be no accumulation of electricity in voltaic combinations, unless the same or similar conditions of chemical change exist in the elements or single circles composing them; and that under other conditions, the power generated in single circles is either destroyed or diminished according to the opposing nature, or want of conducting power of the chain of intervening bodies. For instance, in the arrangement (mentioned p. 331) of one piece of zinc and one of platinum, the power is doubled by another series of the same kind, destroyed by an arc of platinum, and diminished by an arc of zinc; by a second solution and a second arc of zinc, it is diminished still more; by a third it is nearly, and by a fourth absolutely, *destroyed*.

As the chemical changes always tend to restore the electrical equilibrium destroyed by the contact of the metals with each other in the fluids, it is evident that in cases in which arcs primarily inactive are connected with those primarily active, the chemical changes produced by the electrical attractions must tend to produce in the primarily inactive parts of the combination an arrangement which must give it a power in direct opposition to that of the primarily active circles; so that when separated, their actions, if any, must be directly the reverse of the other. This result, which I anticipated, I have actually found to be correct; six arcs of platinum in vessels filled with solution of nitre, were connected with a voltaic battery of fifty pairs of plates; of course each arc gave off oxygen, and collected acid round the pole in the place of the zinc, and afforded hydrogen and collected alkali round the pole in the place of the noble metal: on separating the six arcs from the battery, they were found to possess independent action, the poles which were negative being positive, and those positive being negative: in short, the combination acted as if an original one, consisting of acid, alkali, and platinum.

With arcs of zinc the results were of the same kind, but the electrical effects were much more distinct: as the tarnished zinc in this case added its own negative power to that produced by the contact with the acid.

In trying similar experiments with six arcs of tin, silver, copper and other metals, and using different saline solutions, it was found that the reversed electrical effects were most powerful with the most oxidable metals, and the most concentrated and decomposable solutions; and the weakest arrangement of this kind was with arcs of platinum and pure water; yet even in this instance the water had become slightly alkaline at one pole, and acid at the other.

These experiments, showing the nature of the chemical changes in combinations made active by their connexion with voltaic batteries, and the influence of the newly developed chemical agents, fully explain the phenomena of the secondary piles of M. Ritter; and combined with the fact, that the metals are not PERFECT conductors for electricities of very low intensity, they offer a simple and adequate solution of the circumstances observed by M. De la Rive on the interposition of different metallic plates in the fluids connecting together voltaic combinations.*

From the nature of the chemical changes taking place in each single circle of a common voltaic battery, it is evident, that if any small part of a battery, for some time in action, is separated from the whole, and made to act as a distinct combination, its powers must be feebler than if it had been originally an independent series; for the electrical action occasioned by the chemical agents developed in it, are such as to counteract the effects produced by the contact of the metals. Whereas, if a small voltaic series is connected with a much larger one, in reverse order, its oxidable in the place of the noble metals, though the whole power of the combination is much weakened by it when in union; yet, when separated, it must act with much greater power, as the chemical changes produced are exactly of the kind which must enhance the primary power of the metals. This deduction (a necessary consequence of the electro-chemical theory) I have proved by direct experiment. A series of six arcs, composed of zinc and copper, and solution of nitre, was connected in the proper order with a voltaic arrangement of 50 pairs, and suffered to remain in connexion for 10 minutes; they were then separated, and made to act as a single battery: their powers were extremely feeble, not certainly one-third as great as those of a combination of the same kind which had been in action (but unconnected) for the same time. Six arcs of copper and zinc were now connected with the same battery of 50, in a reverse or unconformable manner, so that the six plates of zinc gave off hydrogen and attracted alkali, and the plates of copper oxidated and attracted acid. Being separated

* Annales de Chimie et de Physique, tom. xxviii. p. 190.

after a few minutes, and made to act alone, they exhibited powers which appeared three or four times greater than if they had never been in connexion; the zinc resumed a much higher positive, and the copper a higher negative state, than if they had not before been in the antagonist or unconformable conditions.

All these facts bear upon the same point, and confirm the view which I took of the nature of voltaic combinations in the Bakerian Lecture for 1806; in all of which, whether the destruction of the electrical equilibrium is produced by the contact of metals or fluids, it is always restored by chemical changes, and in which the circulation, if it may be so called, depends upon a union of these causes, the direction of the currents being always opposite in the metallic and fluid parts of the combination, so as to produce what may be regarded as an electrical circle.

VIII. *General Observations and practical Applications.*

To explain the manner in which different chemical agents in combination, and in a perfectly neutral state, instantly start into an active existence, when exposed to the two electrical poles, it is necessary to assume principles, and take views of corpuscular action of a perfectly novel kind; and as the chief agents are invisible, and probably imponderable, no direct demonstrative evidence can be brought forward on the subject; and different hypotheses may in consequence be applied to it. In assuming the idea of two ethereal, subtile, elastic fluids, attractive of the particles of each other, and repulsive as to their own particles, capable of combining in different proportions with bodies, and according to their proportions giving them their specific qualities and rendering them equivalent masses, it would be natural to refer the action of the poles to the repulsions of the substances combined with excess of one fluid, and the attractions of these united to the excess of the other fluid; and a history of the phenomena, not unsatisfactory to the reason, might in this way be made out; but as it is possible likewise to take an entirely different view of the subject, on the idea of the dependence of the results upon the primary attractive powers of the parts of the combination on a single subtile fluid, I shall not enter into any discussion upon this obscure part of theory, but I shall endeavour to clear the way for elucidations of it by stating some experimental results.

Some solution of nitrate of potassa was introduced into a glass basin of six inches in diameter, and large slips of paper, tinged with litmus and turmeric, were placed below the fluid, and connected with two pieces of foil of platinum; so that the indications of the formation of acid and alkali, in any part of the basin, by electricity, would be instant and distinct. The two pieces of foil were now connected with the poles of a Voltaic battery: it was found that the alkali was developed only at the point or immediate surface of the negative platinum, and the acid in the same manner at the surface of the positive platinum; and they then gradually diffused themselves through the fluid in a circle round the conductors, and there was no appearance of any repulsions or attractions of the menstrua in the line of the circuit.

In various repetitions of this experiment the same result was obtained; the alkaline and acid matters were influenced in their direction only by currents produced by the disengaged oxygen or hydrogen, or the inclination of the vessel; in short, by mechanical causes only: and the same effects were produced on the test papers, as if a spherical piece of acid and an amalgam of potassium had been introduced in the places of the two poles.

Mr. [now Sir John] Herschel has shown, by some elaborate and ingenious experiments in the last Bakerian Lecture,* that an amalgam of potassium, containing so minute a portion as some hundred thousand parts of its weight is strongly attracted so as to occasion violent mechanical motion, by the negative pole in a Voltaic arrangement: and if it be supposed that the fluid is divided into two zones, directly opposite in their powers to the poles of the battery, the virtual change may be regarded as taking place in the two extremities of these zones nearest the neutral point; so that by a series of decompositions and recompositions, the alkaline matters and hydrogen separate at one side, and oxygen, pure or in union, at the other.

In this way, the two electricities may be regarded as the transporters of the ponderable matters, which assume their own peculiar characters at the moment they arrive at the point of rest. I shall detail an experiment which I made under a different form some years ago, and which may assist the imagination in the conception of this singular and mysterious mode of action. A flat glass basin, 10 inches in diameter, was filled with water containing $\frac{1}{2000}$th part of its weight of sulphate of potassa, in the bottom of which 30 or 40 separate globules of mercury, containing from 10 to 100 grains each, were placed without any regard to order; two wires of platinum from a battery of 1000 double plates, weakly charged, were made to connect the extremities

* [Phil. Trans. for 1824.]

of the water (passing to the bottom of the basin.) As soon as the electrical communication was made, the globules of mercury in or near the current became instantly agitated; their negative poles became elongated, and approached either the positive pole of the battery, or the positive pole of the contiguous globules of mercury, and streams of oxide flowed with great rapidity from the positive toward the negative pole. No hydrogen appeared at the negative poles of the globules of mercury; but after the action had continued a few minutes, and was then suspended, there was an appearance of some minute globules, owing, as was proved by tests, to the formation and oxidation of potassium which had combined with the mercury, and which, as is evident from Mr. Herschel's researches, had given to that part of the globule in which it had combined its high electro-positive qualities. When the connexion was again made, the same series of constant and violent motions took place; the elongated and negative extremities of every globule moving towards the positive surfaces, and undergoing continual oscillations; but on pouring a small quantity of muriatic acid into the water, so as to make it slightly acid, these phenomena ceased; the masses of mercury resumed their spherical form, hydrogen was given off from the negative surfaces, and all motion and agitation were at an end. The energy of the acid in this case being negative, may be considered as neutralizing the power of the potassium by its immediate contact, and as destroying all the phenomena of attraction by the positive pole.

In the numerous experiments that I made in 1806, on the transfer of acids to the positive pole and of alkalies to the negative pole, there were similar instances in which masses of acid or alkaline matters, by exerting their own peculiar energies, prevented the accumulation of the antagonist elements at their points of rest, so as to destroy, or materially weaken, their power of motion or transport. For instance, in attempting to transfer baryta from the positive to the negative pole, the negative pole being plunged in sulphuric acid, or sulphuric acid to the positive pole, the negative being plunged in a solution of baryta, the reagents were neutralized, and formed insoluble precipitates at the point of union of the menstrua; and no baryta reached the negative, and no sulphuric acid the positive pole.

With muriatic acid and salts of silver, the case was the same. And when acids and alkalies, forming soluble compounds, were used in similar experiments, a great length of time was required, proportional in some measure to their masses, before a particle of acid reached the positive, or of alkali the negative pole; and the result was not destroyed till after the intermediate combination had taken place to a considerable extent; proving the phenomena of continued decompositions and re-compositions, and showing that the electrical and chemical phenomena are of the same order, and produced by the same cause.

In the Bakerian Lecture for 1806, I proposed the electrical powers, or the forces required to disunite the elements of bodies, as a test or measure of the intensity of chemical union. By the use of the multiplier, it would be now easy to apply this test; and *accurate* researches on the connexion of what may be called the electro-dynamic relations of bodies to their combining masses or proportional numbers, will be the first step towards fixing chemistry on the permanent foundation of the mathematical sciences.

I could enter into some other general views of the pure scientific relations of this subject, and its connexion with thermo-electricity, and the phenomena of cohesion; but having already taken up so much of the time of the Society, I shall defer what I have to say on these subjects to another occasion, and I shall conclude with a few practical observations.

A great variety of experiments made in different parts of the world has proved the full efficacy of the electro-chemical means of preserving metals, particularly the copper sheathing of ships; but a hope I had once indulged, that the peculiar electrical state would prevent the adhesion of weeds or insects, has not been realized; protected ships have often, indeed, returned after long voyages perfectly bright,* and cleaner than unprotected ships, yet this is not always the case: and though the *whole* of the copper may be preserved from chemical solution in steam-vessels by these means, yet they must be adopted in common ships only, so as to preserve a portion—so applied, as to suffer a certain solution of the copper;† and an absolute remedy for adhesions, is to be sought for by other more refined means of protection, and which appear to be indicated by these researches.

The nails used in ships are an alloy of copper and tin,

* The Carnebrea Castle.

† A common cause of adhesions of weeds or shell-fish, is the oxide of iron formed and deposited round the protectors. In the only experiment in which zinc has been employed for this purpose in actual service, the ship returned after two voyages to the West Indies, and one to Quebec, perfectly clean.

The experiment was made by Mr. Lawrence, of Lombard-street, who, in his letter to me, states that the rudder, which was not protected, had corroded in the usual manner.

which I find is slightly negative with respect to copper, and it is on these nails, that the first adhesions uniformly take place: a slightly positive and slightly decomposable alloy would probably prevent this effect, and I have made some experiments favourable to the idea.

In general, all changes in metals which would indicate the power of chemical attraction, are easily determined by electrical means. Thus, I found copper hardened by hammering negative to rolled copper; copper (to use the technical language of manufacturers) both *overpoled* and *underpoled*, containing in one case probably a little charcoal, and in the other a little oxide, negative to pure copper. A specimen of brittle copper, put into my hands by Mr. Vivian, but in which no impurity could be detected, was negative with respect to soft copper.

In general, very minute quantities of the oxidable metals render the alloy positive, unless it becomes harder; in which case it is generally negative. As I have mentioned before, amalgams of the oxidable metals are usually positive, not only to mercury, but even to the pure metals.

There are probably few chemical operations which electrical changes do not influence, and either increase or modify. In the rusting of iron, for instance, the oxide formed by the contact of moisture, becomes the negative surface, and exalts the oxidability of the mass of metallic iron, and the rust consequently extends in a circle.

The precipitations of metals have been already traced to causes of this kind; and many metallic solutions must belong to the same order of phenomena.

I have pointed out, in former papers, some of the

cases of electro-chemical protection, which, I have no doubt, when the principles are well understood, will be generally adopted; and others are constantly occurring. I shall mention one,—the preservation of the iron boilers of steam-engines, by introducing a piece of zinc or tin.* This, in the case of steam-boats, particularly when salt water is used, may be of the greatest advantage, and prevent the danger of explosion, which generally arises from the wear of one part of the boiler.

Another application of importance which may be made, is the prevention of the wear of the paddles or wheels, which are rapidly dissolved by salt water.

But I will conclude. Whenever a principle or discovery involves or unfolds a law of nature, its applications are almost inexhaustible; and however abstracted it may appear, it is sooner or later employed for common purposes of the arts, and the common uses of life.†

* [The use of tin for this purpose has been objected to; it has been found to be inert; zinc is certainly preferable, if a protecting metal be required; but that it is required, appears to be questionable, as iron has not the power of decomposing water at 212°.]

† [A royal medal was awarded the author, by the Council of the Royal Society, for his later electro-chemical researches contained in this, and the preceding papers, on the protection of the copper sheathing of ships.]

XX.

ON THE PHENOMENA OF VOLCANOES.*

When, in the years 1827 and 1828, I discovered that the alkalies and the earths were composed of inflammable matter, united to oxygen, a number of inquiries suggested themselves with respect to various parts of chemical science, some of which were capable of being immediately assisted by experiment, and others required for their solution a long series of observations and circumstances, obtained only with difficulty. Of the last kind were the inferences concerning the geological appearances connected with these discoveries.

The metals of the alkalies, and those of such of the earths as I had decomposed, were found to be highly combustible, and altered by air and water, even at the usual temperatures of the atmosphere; it was not possible, consequently, that they should be found at the surface of the globe, but probable that they might exist in the interior: and allowing this hypothesis, it became easy to account for volcanic fires, by exposure of the metals of earths and alkalies to air and water; and to explain, not only the formation of lavas, but likewise that of basalts, and many other crystalline rocks, from the slow cooling of the products of combustion, or oxidation of the newly discovered substances.

I developed this opinion in a paper on the decompo-

* [From the Phil. Trans. for 1828.—Read before the Royal Society, March 20, 1828.]

sition of the earths, publishedin 1808; and since 1812 I have endeavoured to gain evidence respecting it, by examining volcanic phenomena of ancient and recent occurrence in various parts of Europe.

In this communication I shall have the honour of laying before the Royal Society some results of my inquiries. If they do not solve the problem respecting the cause of volcanic fires, they will, I trust, be found to offer some elucidations of the subject, and may serve as the foundation for future labours.

The active volcano, on which I have made my observations, is Vesuvius; and there probably does not exist another so admirably fitted for the purpose; its vicinity to a great city; the facility with which it may be ascended, in every season of the year; and the nature of its activity,—all offer peculiar advantages to the philosophic inquirer.

I had made several observations on Vesuvius in the springs of 1814 and 1815, which I shall refer to on a future occasion in these pages: but it was in December, 1819, and January and February, 1820, that the volcano offered the most favourable opportunity for investigation. On my arrival at Naples, December 4, I found that there had been a small eruption a few days before, and that a stream of lava was flowing, with considerable activity, from an aperture in the mountain, a little below the crater. On the 5th, I ascended the mountain, and examined the crater and the stream of lava. The crater emitted so large a quantity of smoke, with muriatic and sulphurous acid fumes, that it was impossible to approach it except in the direction of the wind; and it threw up, every two or three minutes, showers of red-hot stones. The lava was flowing out from an aperture about one hundred yards below it, being apparently forced out by elastic fluids, with a noise like that made by the steam disengaged from a pressure engine: it rose perfectly fluid, forming a stream of from five to six feet in diameter, and immediately fell, as a cataract, into a chasm, about forty feet below, where it was lost under a kind of bridge, formed of cooled lava; but it reappeared sixty or seventy yards further down. Where it issued from the mountain, it was nearly white-hot, and exhibited an appearance similar to that which is shown when a pole of wood is introduced into the melted copper of a foundry, its surface appearing in violent agitation, large bubbles rising, which, in bursting, produced a white smoke; but the lava became of a red colour, though still visible in the sunshine, where it issued from under the bridge. The force with which it flowed was so great, that the strength of the guide, a very stout young man, was insufficient to keep a long iron rod in the current. The whole of its course, with two or three interruptions, where it flowed under a cooled surface, was nearly three quarters of a mile, and it threw off clouds of a white smoke. It smoked less as it cooled, and became pasty; but even where it terminated in moving masses of scoria, smoke was still visible, which became more distinct whenever the scoria was moved, or the red-hot lava in the interior exposed.

Having ascertained that it was possible to approach within four or five feet of the lava, and to examine the vapour immediately close to the aperture, I returned the next day, having provided the means of making a number of experiments on the nature of the lava, and of the elastic fluids with which it was accompanied. I found the aperture nearly in the same state as the day before, but the lava spread over a larger surface, forming an eddy in the hollow of the rock, over which it fell, from which it could be raised in an iron ladle more easily than from the current, and where there was much more facility of placing and withdrawing substances intended to be exposed to its agency.

One of the most important points to be ascertained was, whether any combustion was going on at the moment the lava issued from the mountain. There was certainly no appearance of more vivid ignition when it was exposed to air, nor did it glow with more intensity when it was raised into the air in an iron ladle. I put the circumstance, however, beyond the possibility of doubt: I threw some of the fused lava into a glass bottle, furnished with a ground stopper, containing siliceous sand in the bottom: I closed it at the moment, and examined the air on my return. A measure of it, mixed with a measure of nitrous gas, gave exactly the same degree of diminution as a measure of common air, which had been collected in another bottle on the mountain.

I threw upon the surface of the lava nitre, both in mass and in powder. After this salt had fused, there was a little increase of vividness in the ignition of the lava, but much too slight to be referred to pure combustible matter in any quantity; and on making the experiment on a portion of lava taken up in the ladle, it appeared that the disengagement of heat was partly owing to the peroxidation of the protoxide of iron, and to the combination of the alkali of the nitre with the earthy basis of the lava; for where the nitre had melted, the colour had changed from olive to brown. This conclusion was still further proved by the circumstance that chlorate of potash thrown upon the lava did not increase its degree of ignition so much as nitre. When a stick of wood was introduced into a portion of the lava so as to leave a little carbonaceous matter on its surface, nitre or chlorate of potassa then thrown upon it caused it to glow with great brilliancy. Some fused lava was thrown into water, and a glass bottle filled with water held over it to collect the gas disengaged; it was in very minute quantity only, and when analysed on my return proved to be common air a little less pure than that disengaged from the water by boiling. A wire of copper of $\frac{1}{20}$ of an inch in diameter, and a wire of silver of $\frac{1}{30}$, introduced into the lava near its source, were instantly fused: an iron rod of $\frac{1}{5}$ of an inch, with a piece of iron wire of about $\frac{1}{30}$, were kept for five minutes in the eddy in the stream of lava: they were not fused; they did not produce any smell of sulphuretted hydrogen when acted on by muriatic acid. A tin-plate funnel filled with cold water was held in the fumes disengaged with so much violence from the aperture through which the lava issued: fluid was immediately condensed upon it, which was of an acid and astringent taste. It did not precipitate muriate of baryta; but copiously precipitated nitrate of silver, and rendered the triple prussiate of potassa of a bright blue. When the same funnel was held in the white fumes above the lava where it entered the bridge, no fluid was precipitated upon it, but it became coated with a white powder which had the taste and chemical qualities of common salt, and proved to be this substance absolutely pure. A bottle of water holding about $\frac{3}{4}$ of a pint, with a long narrow neck, was emptied immediately in the aperture from which the vapours pressing out the lava issued, and the neck was immediately closed. This air examined on my return was found to give no absorption with solution of potassa, so that it contained no notable proportion of carbonic acid, and it consisted of nine parts of oxygen, and ninety-one of azote. There was not the least smell of sulphurous acid in the vapour from the aperture, nor were the fumes of muriatic acid so strong as to be unpleasant; but during the last quarter of an hour that I was engaged in these experiments, the wind changed and blew the smoke from the crater upon the spot where I was standing: the sulphurous acid gas in the fumes was highly irritating to the organs of respiration, and I suffered so much from the exposure to them that I was obliged to descend; and the effect was not transient, for a violent catarrhal affection ensued, which prevented me for a month from again ascending the mountain.

On the 6th of January I made another visit to Vesuvius. I found the appearance of the lava considerably changed; the bocca from which it issued on the 5th of December was closed, and the current now flowed quietly and without noise from a chasm in the cooled lava about three hundred feet lower down. The heat was evidently less intense. I repeated my experiments with nitre with the same results, and exposed pure silver and platinum to the fused lava: they were not at all changed in colour. I collected the sublimations from various parts of the cooled lava above. The rocks near the ancient bocca were entirely covered with white, yellow, and reddish saline substances. I found one specimen of large saline crystals in a cavity, which had a slight tint of purple: this examined, proved to be common salt with a minute portion of muriate of cobalt. The other sublimations consisted of common salt in great excess, much chloride of iron, some sulphate of soda; and by the test of muriate of platinum, there appeared to exist in them a small quantity of sulphate or muriate of potassa; and a solution of ammonia detected the presence of a minute quantity of the oxide of copper.

During the months of January and February, I made several visits to the top of Vesuvius: I shall not particularize them all; but shall mention only such as afforded me some new observations. On the 26th of January, the lava was seen nearly white-hot through a chasm near the place where it flowed from the mountain. I threw nitre upon it in large quantities through this chasm, in the presence of H.R.H. the Prince of Denmark, whom I had the honour of accompanying in this excursion to the mountain, and my friend the Cavaliere Monticelli: there was no more increase of ignition than when the experiment was made on lava exposed to the free air. The appearance of the sublimations was now considerably changed: those near the aperture were coloured green and blue by salts of copper; but there was still a great quantity of muriate of iron. I have mentioned, that on the 5th the sublimate of the lava was pure chloride of sodium: in the sublimate of January 6th there were both sulphate of soda and indications of sulphate of potassa. In the sublimates that I collected on the 26th, the sulphate of soda was in much larger quantities, and there was much more of a salt of potassa. From the 5th of December to the 20th of February, the lava flowed in larger or smaller quantities, so that at night a stream of ignited matter was always visible, more or less interrupted by cooled lava. It changed its direction according to the obstacles it met with; and never, according to appearances, extended so much as a mile from its source. During the whole of this time, the craters, of which there were two, were in activity. The large crater threw up showers of ignited ashes and stones to a height apparently of from 200 to 500 feet; and from a small crater, to the right of the large one on the side of Naples, steam arose with great violence. Whenever the crater could be approached it was found incrusted with saline incrustations; and the walk to the edge of the small crater on the 6th of January was through a mass of loose saline matter, principally common salt coloured by muriate of iron, in which the foot sunk to some depth. It was easy, even at a great distance, to distinguish between the steam disengaged by one of the craters, and the earthy matter thrown up by the other. The steam appeared white in the day, and formed perfectly white clouds, which reflected the morning and evening light of the purest tints of red and orange. The earthy matter always appeared as a black smoke, forming black clouds; and in the night it was highly luminous at the moment of the explosion.

On the 20th of February, the small crater, which had been disengaging steam and elastic matter, began to throw out showers of stones; and both craters, from the 20th to the 23rd, were more than usually active. On the night of the 23rd, at half-past 11 o'clock, being in my bed-room at Chiatimone, Naples, I heard the windows shake; and, going to the window, I saw ascending from Vesuvius a column of ignited matter, to a height at least equal to that of the mountain from its base; and the whole horizon was illuminated, notwithstanding the brightness of the moon, with direct volcanic light, and that reflected from the clouds above the column of ignited matter. Several eruptions of the same kind, but upon a smaller scale, followed at intervals of a minute and half or two minutes; but there were no more symptoms of earthquake, nor did I hear any noise. On observing the lava, it appeared at its origin much broader and more vivid; and it was evident that a fresh stream had broken out to the right of the former one. On the morning of the 24th, I visited the mountain; it was not possible to ascend to the top, which was covered with clouds,—nor to examine the orifice, from which the lava issued. The stream of lava, near the place where it terminated, was from 50 to 100 feet broad. It had precisely the same appearances as the lava which had been so long running. I collected the saline matter condensed upon some of the masses of scoria, which were carried along by the current, and deposited on the edge of the stream; they proved to be the same, in the nature of their constituent parts, as those of the lava of the 26th of January, but with a larger proportion of sulphate of soda, and a smaller proportion of muriate of iron; and I have no doubt that the dense white smoke, which was emitted in immense columns by the lava, during the whole of its course, was produced by the same substances.

I shall now mention the state of the volcano at some other periods.

When I was at Naples, in May, 1814, the crater had the appearance of an immense funnel, closed at the bottom, with many small apertures, emitting steam; and on the side towards Torre del Greco, there was a large aperture, from which flames issued to a height of at least 60 yards, producing a most violent hissing noise. This phenomenon was constant, during the three weeks I remained at Naples. It was impossible to approach sufficiently near the flame, to ascertain the results of the combustion; but a considerable quantity of steam ascended from it. When the wind blew the vapours upon us, there was a distinct smell, both of sulphurous and muriatic acid. There was no indication of carbonaceous matter, from the colour of the smoke; nor was any deposited upon the yellow and white saline matter, which surrounded the crater, and which I found to be principally sulphate and muriate of soda, and muriate of iron: in some specimens, there was a considerable quantity of muriate of ammonia.

In March, 1815, the appearances presented by the crater, were entirely different. There was no aperture in the crater; it was often quiet for minutes together; and then burst out into explosions, with considerable violence, sending fluid lava and ignited stones and ashes to a considerable height, many hundred feet, in the air.

These eruptions were preceded by subterraneous thunder, which appeared to come from a great distance, and which sometimes lasted for a minute. During the four times that I was upon the crater in the month of March, I had at last learnt to estimate the violence of the eruption, from the nature of the sound: loud and long-continued subterraneous thunder, indicated a considerable explosion. Before the eruption, the crater appeared perfectly tranquil; and the bottom, apparently without an aperture, was covered with ashes. Soon indistinct rumbling sounds were heard, as if at a great distance; gradually the sound approached nearer, and was like the noise of artillery fired under our feet. The ashes then began to rise, and to be thrown out with smoke from the bottom of the crater; and lastly, the lava and ignited matter was ejected, with a most violent explosion. I need not say that when I was standing on the edge of the crater witnessing this phenomenon, the wind was blowing strongly from me; without this circumstance, it would have been dangerous to have stood on the edge of the crater; and whenever, from the loudness of the thunder, the eruptions promised to be violent, I always ran as far as possible from the seat of danger.

As soon as the eruption had taken place, the ashes and stones which rolled down the crater seemed to fill up the aperture, so that it appeared as if the ignited and elastic matter were discharged laterally; and the interior of the crater assumed the same appearance as before.

I shall now offer some observations on the theory of these phenomena. It appears almost demonstrable that none of the chemical causes anciently assigned for volcanic fires, can be true. Amongst these, the combustion of mineral coal is one of the most current: but it seems wholly inadequate to account for the phenomena. However large a stratum of pit-coal, its combustion, under the surface, could never produce violent and extensive heat: for the production of carbonic acid gas, when there was no free circulation of air, must tend constantly to impede the process; and it is scarcely possible that carbonaceous matter, if such a cause existed, should not be found in the lava, and be disengaged with the saline or aqueous products, from the bocca or craters. There are many instances in England of strata of mineral coal, which have been long burning; but the results have been merely baked clay and schists, and it has produced no result similar to lava.

If the idea of Lemery were correct, that the action of sulphur on iron may be a cause of volcanic fires, sulphate of iron ought to be the great product of the volcano; which is known not to be the case; and the heat produced by the action of sulphur on the common metals, is quite inadequate to account for the appearances. When it is considered that volcanic fires occur and intermit with all the phenomena that indicate intense chemical action, it seems not unreasonable to refer them to chemical causes. But for phenomena upon such a scale, an immense mass of matter must be in activity, and the products of the volcano ought to give an idea of the nature of the substances primarily active. Now what are these products? Mixtures of the earths in an oxidated and fused state, and intensely ignited, water and saline substances, such as might be furnished by the sea and air, altered in such a manner as might be expected from the formation of fixed oxidated matter. But it may be said, if the oxidation of the metals of the earths be the causes of the phenomena, some of these substances ought occasionally to be found in the lava, or the combustion ought to be increased at the moment the materials passed into the atmosphere. But the reply to this objection is, that it is evident that the changes which occasion volcanic fires, take place in immense subterranean cavities; and that the access of air to the acting substances occur long before they reach the interior surface.

There is no question but that the ground under the Solfaterra is hollow, and there is scarcely any reason to doubt of a subterranean communication between this crater and that of Vesuvius: whenever Vesuvius is in an active state, the Solfaterra is comparatively tranquil. I examined the bocca of the Solfaterra on the 21st of February, 1820, two days before the activity of Vesuvius was at its height; the columns of steam which usually arise in large quantities when Vesuvius is tranquil, were now scarcely visible; and a piece of paper thrown into the aperture did not rise again; so that there was every reason to suppose the existence of a descending current of air.* The subterraneous thunder heard at such great

* In 1814 and 1815, and in January 1819, when Vesuvius was com-

distances under Vesuvius, is almost a demonstration of the existence of great cavities below filled with aëriform matter; and the same excavations which in the active state of the volcano throw out during so great a length of time immense volumes of steam, must, there is every reason to believe, in its quiet state become filled with atmospheric air.*

To what extent subterranean cavities may exist, even in common rocks, is shown in the limestone caverns of Carniola, some of which contain many hundred thousand cubical feet of air; and in proportion as the depth of an excavation is greater, so is the air more fit for combustion.

The same circumstance which would give alloys of the metals of the earths the power of producing volcanic phenomena, namely their extreme facility of oxidation, must likewise prevent them from being ever found in a pure combustible state in the products of volcanic eruption; for before they reach the external surface, they must not only be exposed to the air in the subterranean cavities, but be propelled by steam; which must possess, under the circumstances, at least the same facility of oxidating them as air.

Assuming the hypothesis of the existence of such alloys of the metals of the earths as may burn into lava in the interior, the whole phenomena may be easily explained from the action of the water of the sea and air on those metals; nor is there any fact, or any of the circumstances which I have mentioned in the preceding part of

paratively tranquil, I observed the Solfaterra in a very active state, throwing up large quantities of steam and some sulphuretted hydrogen.

* Vesuvius is a mountain admirably fitted, from its form and situation, for experiments on the effect of its attraction on the pendulum; and it would be easy in this way to determine the problem of its cavities. On Etna the problem might be solved on a larger scale.

this paper, which cannot be easily explained according to that hypothesis. For almost all the volcanoes in the old world of considerable magnitude are near, or at no considerable distance from the sea; and if it be assumed that the first eruptions are produced by the action of sea-water upon the metals of the earths, and that considerable cavities are left by the oxidated metals thrown out as lava, the results of their action are such as might be anticipated; for after the first eruptions, the oxidations which produce the subsequent ones may take place in the caverns below the surface; and when the sea is distant, as in the volcanoes of South America, they may be supplied with water from great subterranean lakes; as Humboldt states that some of them throw up quantities of fish.

On the hypothesis of a chemical cause for volcanic fires, and reasoning from known facts, there appears to me no other adequate source than the oxidation of the metals which form the bases of the earths and alkalies; but it must not be denied that considerations derived from thermometrical experiments on the temperature of mines and of sources of hot water, render it probable that the interior of the globe possesses a very high temperature; and the hypothesis of the nucleus of the globe being composed of fluid matter, offers a still more simple solution of the phenomena of volcanic fires than that which has been just developed.*

* [Ultimately he gave the preference to the latter hypothesis, affording a strong proof how little he was attached to his own views, merely because they were his own, and what little importance he attached to hypothesis, however brilliant. His latest opinions on this mysterious subject he expressed in his posthumous work, "Consolations in Travel," in the Third Dialogue. The arguments he there uses in favour of the above notion are chiefly those adverted to in the text; the deficiency of evidence in support of the chemical doctrine; the temperature beneath the

Whatever opinion may be ultimately adopted on this subject, I hope that these inquiries on the actual products of a volcano in eruption will not be without interest for the Royal Society.

surface, being found in mines to increase with their depth; and the sources of hot water which have been discovered in all the different rock formations, and so profusely scattered. The negative evidence considered without bias appears very strong and almost irresistible; especially the absence of any considerable quantity of hydrogen amongst volcanic products; and also the absence of metallic iron; and in hot springs the absence in a notable quantity of the soluble earths or alkalies, or of any other substances, the bases of which by union with oxygen could produce heat.]

XXI.

AN ACCOUNT OF SOME EXPERIMENTS ON THE TORPEDO.*

Amidst the variety of researches which have been pursued respecting the different forms and modes of excitation and action of electricity, it is surprising to me that the electricity of living animals has not been more an object of attention, both on account of its physiological importance, and its general relation to the science of electro-chemistry.

In reading an account of the experiments of Walsh, it is impossible not to be struck by some peculiarities of the electricity of the organ of the Torpedo and Gymnotus; such as its want of power to pass through air, and the slight effects of ignition produced by the strongest shocks: and though Mr. Cavendish, with his usual sagacity, compared its action to that of a battery weakly charged, when the electricity was large in quantity but low in intensity, yet the peculiarities

[* From Phil. Trans. for 1829.—Read before the Royal Society, Nov. 20, 1828.]

[Though all the results in this paper are of a negative kind, and many of them not in accordance with after experiments, as has been noticed in the proper place in the First Volume, I apprehend it will not be read without interest, or be considered without value;—partly on account of the philosophical tone in which it is written; partly, as it was the author's last contribution to the Royal Society; and in part, on account of the very distinct manner in which it marks the boundaries of the field of research to which it belongs, as then explored.]

which I have just mentioned are not entirely in harmony with this view of the subject.

When Volta discovered his wonderful pile, he imagined he had made a perfect resemblance of the organ of the Gymnotus and Torpedo; and whoever has felt the shocks of the natural and artificial instruments, must have been convinced, as far as sensation is concerned, of their strict analogy. After the discovery of the chemical power of the voltaic instrument, I was desirous of ascertaining if this property of electricity was possessed by the electrical organs of living animals; and being in 1814 and 1815 on the coast of the Mediterranean, I made use of the opportunities which offered themselves of making experiments on this subject. Having obtained in the Bay of Naples, in May 1815, two small torpedos alive, I passed the shocks through the interrupted circuit made by silver wire, without being able to perceive the slighest decomposition of that fluid; and I repeated the same experiments at Mola di Gaeta, with an apparatus in which the smallest possible surface of silver was exposed, and in which good conductors, such as solutions of potassa and sulphuric acid, were made to connect the circuit; but with the same negative results.

Having obtained a larger Torpedo at Rimini, in June in the same year, I repeated the experiments, using all the precautions I could imagine, with the like results; and at the same time I passed the shock through a very small circuit, which was completed by a quarter of an inch of extremely fine silver wire, drawn by the late Mr. Cavendish for using in a micrometer, and which was less than $\frac{1}{1000}$th of an inch in diameter; but no ignition of the wire took place. It appeared to me after these experiments, that the comparison of the organ of the Torpedo to an electrical battery weakly charged, and of which the charged surfaces were imperfect conductors, such as water, was more correct than that of the comparison to the pile: but on mentioning my researches to Signor Volta, with whom I passed some time at Milan that summer, he showed me another form of his instrument, which appeared to him to fulfil the conditions of the organs of the torpedo; a pile, of which the fluid substance was a very imperfect conductor, such as honey or a strong saccharine extract, which required a certain time to become charged, and which did not decompose water, though when charged it communicated weak shocks.

The discovery of Œrsted, of the effects of voltaic electricity on the magnetic needle, made me desirous to ascertain if the electricity of living animals possessed this power; and after several vain attempts to procure living torpedos sufficiently strong and vigorous to give powerful shocks, I succeeded in October of this year, through the kind assistance of George Deering, Esq., His Majesty's Consul at Trieste, in obtaining two lively and recently caught torpedos, one a foot long, the other smaller. I passed the shocks from the largest of these animals a number of times through the circuit of an extremely delicate magnetic electrometer, (of the same kind, but more sensible than that I have described in my last paper on the electro-chemical phenomena, which the Royal Society has honoured with a place in their Transactions for 1826,) but without perceiving the slightest deviation of an effect on the needle; and I convinced myself that the circuit was perfect, by making my body several times a part of it, holding the silver spoon, by which the shock was taken, in one hand, wetted in salt and water, and keeping the wire connected with the electrometer in the other wet hand; the shocks which passed through the reduplications of the electrometer were sufficiently powerful to be felt at both elbows, and once even in the shoulders.

The negative results may be explained by supposing that the motion of the electricity in the torpedinal organ is in no measurable time, and that a current of some continuance is necessary to produce the deviation of the magnetic needle; and I found that the magnetic electrometer was equally insensible to the weak discharge of a Leyden jar as to that of the torpedinal organ; though whenever there was a continuous current from the smallest surfaces in voltaic combinations of the weakest power, but in which some chemical action was going on, it was instantly and powerfully affected. Two series of zinc and silver, and paper moistened in a lt and water, caused the permanent deviation of the needle several degrees, though the plates of zinc were only $\frac{1}{6}$ of an inch in diameter.

It would be desirable to pursue these inquiries with the electricity of the Gymnotus, which is so much more powerful than that of the Torpedo: but if they are now to be reasoned upon, they seem to show a stronger analogy between common and animal electricity, than between voltaic and animal electricity: it is, however, I think, more probable that animal electricity will be found of a distinctive and peculiar kind.

Common electricity is excited upon non-conductors and readily carried off by conductors and imperfect conductors. Voltaic electricity is excited upon combinations of perfect and imperfect conductors, and is only transmitted by perfect conductors, or imperfect conductors of the best kind.

Magnetism, if it be a form of electricity, belongs only to perfect conductors; and, in its modifications, to a peculiar class of them.

The animal electricity resides only in the imperfect conductors forming the organs of living animals, and its object in the economy of nature is to act on living animals.

Distinctions might be established in pursuing the various modifications or properties of electricity in these different forms, but it is scarcely possible to avoid being struck by another relation of this subject. The torpedinal organ depends for its power upon the will of the animal. John Hunter has shown how copiously it is furnished with nerves. In examining the columnar structure of the organ of the Torpedo, I have never been able to discover arrangements of different conductors similar to those in galvanic combinations, and it seems not improbable that the shock depends upon some property developed by the action of the nerves.

To attempt to reason upon any phenomena of this kind as dependent upon a specific fluid, would be wholly vain.

Little as we know of the nature of electrical action, we are still more ignorant of the nature and of the functions of the nerves. There seems, however, a gleam of light worth pursuing in the peculiarities of animal electricity, its connection with so large a nervous system, its dependence upon the will of the animal, and the instantaneous nature of its transfer, which may lead, when pursued by adequate inquiriers, to results important for physiology.

The weak state of my health will, I fear, prevent me from following this subject with the attention it seems to desire; and I communicate these imperfect trials to the Royal Society, in the hope that they may lead to more extensive and profound researches.

October 24th, 1828,
Lubiana, Illyria.

END OF VOL. VI.

PRINTED BY STEWART AND MURRAY, OLD BAILEY.

Printed in Dunstable, United Kingdom

91219935R00221